MONOGRAPHS ON
STATISTICS AND APPLIED PROBABILITY

General Editors

D.R. Cox, D.V. Hinkley, D. Rubin and B.W. Silverman

(Full details concerning this series are available from the Publishers.)

Monte Carlo Methods

J. M. Hammersley
*Oxford University Institute of
Economics and Statistics*

and

D. C. Handscomb
Oxford University Computing Laboratory

CHAPMAN & HALL
London · Glasgow · New York · Tokyo · Melbourne · Madras

Published by Chapman & Hall, 2–6 Boundary Row, London SE1 8HN

Chapman & Hall, 2–6 Boundary Row, London SE1 8HN, UK

Blackie Academic & Professional, Wester Cleddens Road, Bishopbriggs, Glasgow G64 2NZ, UK

Chapman & Hall, 29 West 35th Street, New York NY10001, USA

Chapman & Hall Japan, Thomson Publishing Japan, Hirakawacho Nemoto Building, 6F, 1-7-11 Hirakawa-cho, Chiyoda-ku, Tokyo 102, Japan

Chapman & Hall Australia, Thomas Nelson Australia, 102 Dodds Street, South Melbourne, Victoria 3205, Australia

Chapman & Hall India, R. Seshadri, 32 Second Main Road, CIT East, Madras 600 035, India

First edition 1964
Reprinted 1965 (twice), 1979, 1983, 1992

© 1964 J. M. Hammersley and D. C. Handscomb

Printed in Great Britain by The Ipswich Book Company Ltd

ISBN 0 412 15870 1

Contents

Authors' Preface

This monograph surveys the present state of Monte Carlo methods. Although we have dallied with certain topics that have interested us personally, we hope that our coverage of the subject is reasonably complete; at least we believe that this book and the references in it come near to exhausting the present range of the subject. On the other hand, there are many loose ends; for example we mention various ideas for variance reduction that have never been seriously applied in practice. This is inevitable, and typical of a subject that has remained in its infancy for twenty years or more. We are convinced nevertheless that Monte Carlo methods will one day reach an impressive maturity.

The main theoretical content of this book is in Chapter 5; some readers may like to begin with this chapter, referring back to Chapters 2 and 3 when necessary. Chapters 7 to 12 deal with applications of the Monte Carlo method in various fields, and can be read in any order. For the sake of completeness, we cast a very brief glance in Chapter 4 at the direct simulation used in industrial and operational research, where the very simplest Monte Carlo techniques are usually sufficient.

We assume that the reader has what might roughly be described as a 'graduate' knowledge of mathematics. The actual mathematical techniques are, with few exceptions, quite elementary, but we have freely used vectors, matrices, and similar mathematical language for the sake of conciseness.

J. M. HAMMERSLEY
D. C. HANDSCOMB

CHAPTER 1

The General Nature of Monte Carlo Methods

1.1 Mathematical theory and experiment

We often classify mathematicians as either pure or applied; but there are, of course, many other ways of cataloguing them, and the fashions change as the character of mathematics does. A relatively recent dichotomy contrasts the theoretical mathematician with the experimental mathematician. These designations are like those commonly used for theoretical and experimental physicists, say; they are independent of whether the objectives are pure or applied, and they do not presuppose that the theoretician sits in a bare room before a blank sheet of paper while the experimentalist fiddles with expensive apparatus in a laboratory. Although certain complicated mathematical experiments demand electronic computers, others call for no more than paper and pencil. The essential difference is that theoreticians deduce conclusions from postulates, whereas experimentalists infer conclusions from observations. It is the difference between deduction and induction.

For those who care to hunt around and stretch a point, experimental mathematics is as old as the hills. The Old Testament (1 Kings vii. 23 and 2 Chronicles iv. 2) refers to an early mathematical experiment on the numerical value of the constant π. In this case the apparatus consisted of King Solomon's temple, and the experimenter observed its columns to be about three times as great in girth as breadth. It would be nice to think that he inferred that this was a universal property of circular objects; but the text does not say so. On the other hand, experimental mathematics on anything like a substantial scale is quite a recent innovation, particularly if confined to experiments on mathematical objects, such as numbers or equations or polygons.

Monte Carlo methods comprise that branch of experimental mathematics which is concerned with experiments on random numbers. In the last decade they have found extensive use in the fields of operational research and nuclear physics, where there are a variety of problems beyond the available resources of theoretical mathematics. They are still in an early stage of development; but, even so, they have been employed sporadically in numerous other fields of science, including chemistry, biology, and medicine.

Problems handled by Monte Carlo methods are of two types called probabilistic or deterministic according to whether or not they are directly concerned with the behaviour and outcome of random processes. In the case of a probabilistic problem the simplest Monte Carlo approach is to observe random numbers, chosen in such a way that they directly simulate the physical random processes of the original problem, and to infer the desired solution from the behaviour of these random numbers. Thus we may wish to study the growth of an insect population on the basis of certain assumed vital statistics of survival and reproduction. If analysis fails us, we can at least set up a model with paper entries for the life histories of individual insects. To each such individual we may allot random numbers for its age at the births of its offspring and at its death; and then treat these and succeeding offspring likewise. Handling the random numbers to match the vital statistics, we get what amounts to a random sample from the population, which we can analyse as though it were data collected by the entomologist in the laboratory or in the field. But the artificial data may suit us better if easier to amass, or if it lets us vary the vital statistics to an extent that nature will not permit. This sort of Monte Carlo work need not call for electronic computers: Leslie and Chitty [1]† made a Monte Carlo study of capture-recapture biometrics using no more tools than a tinful of Lotto bits. Design studies of nuclear reactors and of telephone exchanges provide other examples of probabilistic problems. The fundamental particles of nuclear physics seem to obey probabilistic rather than deterministic laws. Hence we can simulate the perform-

† Numbers in square brackets refer to the bibliography on pp. 150–158.

ance of a nuclear reactor by choosing random numbers which represent the random motions of the neutrons in it. In this way we can experiment with the reactor without incurring the cost, in money, time, and safety, of its actual physical construction. If the geometry of the reactor is at all complicated, which it usually is in practice, one will probably need large-scale computing equipment to trace out the life-histories of individual neutrons in accordance with the random numbers governing them. 'In telephone systems' writes Thomson [2], 'the dependence on chance arises because, so far as the planner is concerned, the demand for service at any one time, dependent as it is on the individual decisions of a large number of subscribers, is quite unpredictable.... The planner works in terms of an average calling rate corresponding to the busy hour of the day, with random fluctuations described by some suitable probability distribution, and he has to decide how many of the individual switches and so on that are used in the course of a telephone call should be installed.... In the early days, there was no theoretical analysis available; as time went on, theory improved, but systems became more complicated and there will always be practical questions which cannot be readily answered from the existing theory.' And Thomson goes on to describe how the British Post Office built an analogue computer in 1949 to simulate the random fluctuations of telephone traffic and to analyse the resulting Monte Carlo experiment.

One of the main strengths of theoretical mathematics is its concern with abstraction and generality: one can write down symbolic expressions or formal equations which abstract the essence of a problem and reveal its underlying structure. However, this same strength carries with it an inherent weakness: the more general and formal its language, the less is theory ready to provide a numerical solution in a particular application. The idea behind the Monte Carlo approach to deterministic problems is to exploit this strength of theoretical mathematics while avoiding its associated weakness by replacing theory by experiment whenever the former falters. Specifically, suppose we have a deterministic problem which we can formulate in theoretical language but cannot solve by theoretical means. Being deterministic, this problem has no direct association with random processes; but, when theory has exposed its underlying structure, we may perhaps recognize that this structure or formal expression also describes some apparently unrelated random process, and hence we can solve the deterministic problem numerically by a Monte Carlo simulation of this concomitant probabilistic problem. For example, a problem in electromagnetic theory may require the solution of Laplace's equation subject to certain boundary conditions which defeat standard analytical techniques. Now Laplace's equation occurs very widely and, *inter alia*, in the study of particles which diffuse randomly in a region bounded by absorbing barriers. Thus we can solve the electromagnetic problem by performing an experiment, in which we guide the particles by means of random numbers until they are absorbed on barriers specially chosen to represent the prescribed boundary conditions.

This technique of solving a given problem by a Monte Carlo simulation of a different problem has sometimes been called sophisticated Monte Carlo, to distinguish it from straightforward simulation of the original problem. There are various degrees of sophistication: for instance, one may start from a given probabilistic problem, formulate it in theoretical terms, discern a second probabilistic problem described by the resulting theory, and finally solve the first probabilistic problem by simulating the second. The second problem may be a greater or a lesser distortion of the first, or it may even be totally different in character: the only thing that matters is that it shall have the same numerical solution as the first, or more generally that the wanted parts of the two solutions shall differ by a negligible amount, there being no need to ensure agreement between unwanted parts of the two solutions.

There are various reasons for indulging in sophisticated Monte Carlo methods. The main reason springs from the inferential nature of Monte Carlo work. Whenever one is inferring general laws on the basis of particular observations associated with them, the conclusions are uncertain inasmuch as the particular observations are only a more or less representative sample from the totality of all observations which might have been made. Good experimentation tries to ensure that the sample shall be more rather than less representative; and good presentation of the conclusions indicates how likely they are to be wrong by how much. Monte Carlo answers are uncertain because they arise from raw observational data consisting of random numbers; but they can nevertheless serve a useful purpose if we can manage to make the uncertainty fairly negligible, that is to say to make it unlikely that the answers are wrong by very much. Perhaps it is worth labouring this point a little, because some people feel genuine distress at the idea of a mathematical result which is necessarily not absolutely certain. Applied mathematics is not in any case a black-and-white subject; even in theoretical applied mathematics there is always doubt whether the postulates are adequately valid for the physical situation under consideration, and whenever in a theoretical formula one substitutes the value of some experimentally determined quantity, such as the velocity of light or the constant of gravitation, one gets an uncertain numerical result. But this is no cause for worry if the uncertainty is negligible for practical purposes.

One way of reducing uncertainty in an answer is to collect and base it upon more observations. But often this is not a very economic course of action. Broadly speaking, there is a square law relationship between the error in an answer and the requisite number of observations; to reduce it tenfold calls for a hundredfold increase in the observations, and so on. To escape a formidable or even impracticable amount of experimental labour, it is profitable to change or at least distort the original problem in such a way that the uncertainty in the answers is reduced. Such procedures are known as variance-reducing techniques, because uncertainty can be measured in terms of a quantity called variance. In this direction the mathematical experimentalist is more fortunate than the experimentalist in the physical sciences: his experimental material consists of mathematical objects which can be distorted, controlled, and modified more easily and to a much greater extent than material subject to physical limitations, such as instrumental errors or structural tolerances or phenomena affected by weather and climate and so on.

Although the basic procedure of the Monte Carlo method is the manipulation of random numbers, these should not be employed prodigally. Each random number is a potential source of added uncertainty in the final result, and it will usually pay to scrutinize each part of a Monte Carlo experiment to see whether that part cannot be replaced by exact theoretical analysis contributing no uncertainty.

Moreover, as experimental work provides growing insight into the nature of a problem and suggests appropriate theory, good Monte Carlo practice may be to this extent self-liquidating [3].

Experimental mathematicians have not come to replace theoretical ones. Each relies on the other, as with other scientists. The experimenter needs theory to give structure and purpose to his experiments; and the theoretician needs experiment to assess the value and position of his theory. The Monte Carlo experimentalist needs wide experience of formulae and results in pure mathematics, and especially the theory of probability, in order that he may discern those connexions between apparently dissimilar problems which suggest sophisticated Monte Carlo methods. He has to exercise ingenuity in distorting and modifying problems in the pursuit of variance-reducing techniques. He has to be competent at statistical and inferential procedures in order to extract the most reliable conclusions from his observational data. The tools of his trade include computing machinery, and he must be familiar with numerical analysis. As in all experimental work, a feel for the problem is a great and sometimes essential asset. Finally Monte Carlo methods are recent innovations still under development, general devices are few in number, and a great deal depends upon having enough originality to create special methods to suit individual problems. Despite all these demands, Monte Carlo work is a subject with which some acquaintance is well worth while for anyone who has to deal with mathematical problems encountered in real life, as opposed to ones engineered to exemplify textbook theory. For real-life problems mathematical experiment is a most necessary alternative to theory. In short, Monte Carlo methods constitute a fascinating, exacting, and often indispensable craft with a range of applications that is already very wide yet far from fully explored.

1.2 Brief history of Monte Carlo methods

The name and the systematic development of Monte Carlo methods dates from about 1944. There are however a number of isolated and undeveloped instances on much earlier occasions. For example, in the second half of the nineteenth century a number of people performed experiments, in which they threw a needle in a haphazard manner onto a board ruled with parallel straight lines and inferred the value of π from observations of the number of intersections between needle and lines. An account of this playful diversion (indulged in by a certain Captain Fox, amongst others, whilst recovering from wounds incurred in the American Civil War) occurs in a paper by Hall [4]. In the early part of the twentieth century, British statistical schools indulged in a fair amount of unsophisticated Monte Carlo work; but most of this seems to have been of a didactic character and rarely used for research or discovery. The belief was that students could not really appreciate the consequences of statistical theory unless they had seen it exemplified with the aid of laboratory apparatus: demonstrators therefore poured lead shot down boards studded with pins, which deflected the shot in a random fashion into several collecting boxes, and the students were required to see that the frequency of the shot in the various boxes conformed more or less to the predictions of theory; they drew numbered counters from jam-jars and pots (called urns for the sake of scientific dignity) and verified that averages of various sets of such numbers behaved as sampling theory said they should, and so on. Only on a few rare occasions was the emphasis on original discovery rather than comforting verification. In 1908 Student (W. S. Gosset) used experimental sampling to help him towards his discovery of the distribution of the correlation coefficient. Apparently he knew some of the moments of the distribution and had conjectured, perhaps on the basis of these or perhaps by way of Occam's razor, that the analytical form would be proportional to $(1 - \alpha r^2)^\beta$, one of Pearson's frequency curves, where r is the correlation coefficient and α and β are constants depending upon the sample size n. Having fitted samples for $n = 4$ and $n = 8$ to this conjectured expression, and having rounded the resulting estimates of α and β, he guessed that $\alpha = 1$ and $\beta = \frac{1}{2}(n-4)$, which happens to be the exact theoretical result. This remarkable achievement is nevertheless rather different from, though in a sense better than, the ordinary use of Monte Carlo methods nowadays, in which there is little attempt to guess exact results. In the same year Student also used sampling to bolster his faith in his so-called t-distribution, which he had derived by a somewhat shaky and incomplete theoretical analysis.

One consequence of this didactic and verifying rôle for sampling experiments was that the experiments were deliberately shorn of distracting improvements, which might have been employed to sharpen the accuracy of the results: statisticians were insistent that other experimentalists should design experiments to be as little subject to unwanted error as possible, and had indeed given important and useful help to the experimentalist in this way; but in their own experiments they were singularly inefficient, nay, negligent in this respect.

The real use of Monte Carlo methods as a research tool stems from work on the atomic bomb during the second world war. This work involved a direct simulation of the probabilistic problems concerned with random neutron diffusion in fissile material; but even at an early stage of these investigations, von Neumann and Ulam refined this direct simulation with certain variance-reducing techniques, in particular 'Russian roulette' and 'splitting' methods [5]. However, the systematic development of these ideas had to await the work of Harris and Herman Kahn in 1948.

The possibility of applying Monte Carlo methods to deterministic problems was noticed by Fermi, von Neumann, and Ulam and popularized by them in the immediate post-war years. About 1948 Fermi, Metropolis, and Ulam obtained Monte Carlo estimates for the eigenvalues of the Schrödinger equation. Dr Stephen Brush (of the Radiation Laboratory at Livermore), who has a particular interest in the history of mathematics, has unearthed a paper by Kelvin [6] in which sixty years ago astonishingly modern Monte Carlo techniques appear in a discussion of the Boltzmann equation. But Lord Kelvin was more concerned with his results than with his (to him, no doubt, obvious) methods, and it seems entirely right and proper that Ulam, von Neumann, and Fermi should take the credit for not only independently rediscovering Monte Carlo methods but also ensuring that their scientific colleagues should become aware of the possibilities, potentialities, and physical applications. The dissemination of ideas is always an essential component of their production, and never more so than in today's conditions of prolific discovery and publication.

The ensuing intensive study of Monte Carlo methods in the 1950s, particularly in the U.S.A., served paradoxically enough to discredit the subject. There was an understandable attempt to solve every problem in sight by Monte Carlo, but not enough attention paid to which of these problems it could solve efficiently and which it could only handle inefficiently; and proponents of conventional numerical methods were not above pointing to those problems where Monte Carlo methods were markedly inferior to numerical analysis. Their case is weakened by their reluctance to discuss advanced techniques, as Morton remarks [7].

In the last few years Monte Carlo methods have come back into favour. This is mainly due to better recognition of those problems in which it is the best, and sometimes the only, available technique. Such problems have grown in number, partly because improved variance-reducing techniques recently discovered have made Monte Carlo efficient where it had previously been inefficient, and partly because Monte Carlo methods tend to flourish on problems that involve a mass of practical complications of the sort encountered more and more frequently as applied mathematics and operational research come to grips with actualities.

CHAPTER 2

Short Resumé of Statistical Terms

2.1 Object

Being inferential in character, Monte Carlo methods rely on statistics; but at the same time they are useful in several fields where statistical jargon is unfamiliar. Some readers may therefore find it convenient to have a brief account of those statistical techniques and terms which arise most often in Monte Carlo work. What follows in this chapter is a mere outline; fuller information appears in standard textbooks such as Cochran [1], Cramér [2], Kendall and Stuart [3], and Plackett [4]. Although Markov chains are important in Monte Carlo work, we defer a discussion of them until §9.1.

2.2 Random events and probability

A *random event* is an event which has a chance of happening, and *probability* is a numerical measure of that chance. Probability is a number lying between 0 and 1, both inclusive; higher values indicate greater chances. An event with zero probability (effectively) never occurs; one with unit probability surely does. We write: $P(A)$ for the probability that an event A occurs; $P(A+B+\ldots)$ for the probability that at least one of the events A, B, ... occurs; $P(AB\ldots)$ for the probability that all the events A, B, ... occur; and $P(A|B)$ for the probability that the event A occurs when it is known that the event B occurs. $P(A|B)$ is called the *conditional probability of A given B*. The two most important axioms which govern probability are

$$P(A+B+\ldots) \leqslant P(A)+P(B)+\ldots, \qquad (2.2.1)$$

and
$$P(AB) = P(A|B)P(B). \qquad (2.2.2)$$

If only one of the events A, B, ... can occur, they are called *exclusive*, and equality holds in (2.2.1). If at least one of the events $A, B, \ldots$ must occur, they are called *exhaustive*, and the left-hand side of (2.2.1) is 1. If $P(A|B) = P(A)$, we say that A and B are *independent*: effectively, the chance of A occurring is uninfluenced by the occurrence of B.

2.3 Random variables, distributions, and expectations

Consider a set of exhaustive and exclusive events, each characterized by a number η. The number η is called a *random variable*; and with it is associated a *cumulative distribution function* $F(y)$, defined to be the probability that the event which occurs has a value η not exceeding a prescribed y. This may be written

$$F(y) = P(\eta \leqslant y). \tag{2.3.1}$$

The adjective 'cumulative' is often omitted. Clearly $F(-\infty) = 0$, $F(+\infty) = 1$, and $F(y)$ is a non-decreasing function of y.

If $g(\eta)$ is a function of η, the *expectation* (or *mean value*) of g is denoted and defined by

$$\mathscr{E}g(\eta) = \int g(y)\,dF(y). \tag{2.3.2}$$

The integral here is taken over all values of y. For full generality (2.3.2) should be interpreted as a Stieltjes integral; but those who care little for Stieltjes integrals will not lose much by interpreting (2.3.2) in one of the two following senses: (i) if $F(y)$ has a derivative $f(y)$, take (2.3.2) to be

$$\mathscr{E}g(\eta) = \int g(y)f(y)\,dy; \tag{2.3.3}$$

(ii) if $F(y)$ is a step-function with steps of height f_i at the points y_i, take (2.3.2) to be

$$\mathscr{E}g(\eta) = \sum_i g(y_i)f_i. \tag{2.3.4}$$

The point of the Stieltjes integral is to combine (2.3.3) and (2.3.4) in a single formula, and also to include certain possibilities not covered by (i) or (ii). For conciseness we use the form (2.3.2) in this book; and the reader may, if he likes, interpret it as (2.3.3) or (2.3.4) according to context. Effectively, the expectation of $g(\eta)$ is the weighted average of $g(\eta)$, the weights being the respective probabilities of different possible values of η.

Sometimes it is convenient to characterize exhaustive and exclusive events with a vector $\boldsymbol{\eta}$ (i.e. a set of numbers, called the co-ordinates of $\boldsymbol{\eta}$). This gives rise to a *vector random variable*, and an associated distribution function

$$F(\mathbf{y}) = P(\boldsymbol{\eta} \leqslant \mathbf{y}) \qquad (2.3.5)$$

where $\boldsymbol{\eta} \leqslant \mathbf{y}$ means that each co-ordinate of $\boldsymbol{\eta}$ is not greater than the corresponding co-ordinate of $\mathbf{y}$. As before, we define expectations by

$$\mathscr{E}\mathbf{g}(\boldsymbol{\eta}) = \int \mathbf{g}(\mathbf{y}) \, dF(\mathbf{y}). \qquad (2.3.6)$$

The interpretation of (2.3.6) in the sense of (2.3.3) is

$$\mathscr{E}\mathbf{g}(\mathbf{y}) = \int \mathbf{g}(\mathbf{y}) f(\mathbf{y}) \, d\mathbf{y}, \qquad (2.3.7)$$

where, if $F(\mathbf{y}) = F(y_1, y_2, \ldots, y_k)$, we write

$$f(\mathbf{y}) = f(y_1, y_2, \ldots, y_k) = \frac{\partial^k F(y_1, y_2, \ldots, y_k)}{\partial y_1 \, \partial y_2 \ldots \partial y_k},$$

and $\qquad\qquad\qquad\qquad\qquad\qquad\qquad\qquad\qquad$ (2.3.8)

$$d\mathbf{y} = dy_1 \, dy_2 \ldots dy_k.$$

Similarly there is a form corresponding to (2.3.4). The notation $d\mathbf{y}$ used in (2.3.8) is particularly convenient and will be often used in this book. The integrals (2.3.6) and (2.3.7) are multi-dimensional and taken over all values of $y_1, y_2, \ldots$. We normally use lower case **bold** type for vectors and upper case bold type for matrices. We also write $\mathbf{y}'$ and $\mathbf{V}'$ for the transposes of $\mathbf{y}$ and $\mathbf{V}$, and $|\mathbf{V}|$ for the determinant of $\mathbf{V}$.

The quantities $f(y)$ and f_i, appearing in (2.3.3) and (2.3.4), are called the *frequency functions* of the random variable η. Sometimes $f(y)$ is called a *probability density function*.

Consider a set of exhaustive and exclusive events, each characterized by a pair of numbers η and ζ, for which $F(y, z)$ is the distribution function. From this given set of events, we can form a new set, each event of the new set being the aggregate of all events in the old set which have a prescribed value of η. The new set will have a distribution function $G(y)$, say. Similarly we have a distribution function $H(z)$ for the random variable ζ. Symbolically

$$F(y,z) = P(\eta \leqslant y, \zeta \leqslant z); \quad G(y) = P(\eta \leqslant y); \quad H(z) = P(\zeta \leqslant z).$$
(2.3.9)

If it so happens that

$$F(y,z) = G(y) H(z) \text{ for all } y \text{ and } z, \qquad (2.3.10)$$

the random variables η and ζ are called *independent*. Effectively, knowledge of one is no help in predicting the behaviour of the other. This idea of independence can be extended to several random variables, which may be vectors:

$$\begin{aligned} F(\mathbf{y}_1, \mathbf{y}_2, \ldots, \mathbf{y}_k) &= P(\mathbf{\eta}_1 \leqslant \mathbf{y}_1, \mathbf{\eta}_2 \leqslant \mathbf{y}_2, \ldots, \mathbf{\eta}_k \leqslant \mathbf{y}_k), \\ &= P(\mathbf{\eta}_1 \leqslant \mathbf{y}_1) P(\mathbf{\eta}_2 \leqslant \mathbf{y}_2) \ldots P(\mathbf{\eta}_k \leqslant \mathbf{y}_k) \\ &= F(\mathbf{y}_1) F(\mathbf{y}_2) \ldots F(\mathbf{y}_k). \end{aligned} \qquad (2.3.11)$$

Random variables which are independent in pairs are not necessarily mutually independent as in (2.3.11).

The definition of expectation implies that

$$\sum_i \mathscr{E} g_i(\eta_i) = \mathscr{E} \sum_i g_i(\eta_i) \qquad (2.3.12)$$

whether or not the random variables η_i are independent. On the other hand the equation

$$\prod_i \mathscr{E} g_i(\eta_i) = \mathscr{E} \prod_i g_i(\eta_i) \qquad (2.3.13)$$

is true for independent η_i, though generally false for dependent η_i. Equation (2.3.12) is of great importance in Monte Carlo work. As a caution, it should be added that the relation $\mathscr{E} g(\eta) = g(\mathscr{E} \eta)$ is rarely true. Equations (2.3.12) and (2.3.13) also hold for vector random variables.

The quantity $\mathscr{E}(\eta^r)$ is called the rth *moment* of η. Similarly the quantity $\mu_r = \mathscr{E}\{(\eta - \mu)^r\}$, where $\mu = \mathscr{E}\eta$, is called the rth *central moment* of η. By far the most important moments are μ, known as the *mean* of η; and μ_2, known as the *variance* of η. The mean is a measure of location of a random variable, whereas the variance is a measure of dispersion about that mean. The *standard deviation* is defined by $\sigma = \sqrt{\mu_2}$. The *coefficient of variation* is σ divided by μ. It is sometimes expressed as a percentage.

If η and ζ are random variables with means μ and ν respectively, the quantity $\mathscr{E}\{(\eta-\mu)(\zeta-\nu)\}$ is called the *covariance* of η and ζ. Notice that, by (2.3.13), the covariance vanishes if η and ζ are independent; though the converse is generally false. We use the abbreviations *var* and *cov* for variance and covariance. Notice that $\mathrm{cov}(\eta,\eta) = \mathrm{var}\,\eta$. The *correlation coefficient* between η and ζ is defined as $\rho = \mathrm{cov}(\eta,\zeta)/\sqrt{(\mathrm{var}\,\eta\,\mathrm{var}\,\zeta)}$. It always lies between ± 1. If $\rho = 0$, then η and ζ are said to be *uncorrelated*: they are *positively correlated* if $\rho > 0$ and *negatively correlated* if $\rho < 0$.

The above definitions yield the important formula

$$\mathrm{var}\left(\sum_{i=1}^{k} \eta_i\right) = \sum_{i=1}^{k}\sum_{j=1}^{k} \mathrm{cov}(\eta_i,\eta_j). \qquad (2.3.14)$$

The following approximate formula is useful in applications:

$$\mathrm{var}\,g(\eta_1,\eta_2,\ldots,\eta_k) \simeq \sum_{i=1}^{k}\sum_{j=1}^{k} g_i g_j \mathrm{cov}(\eta_i,\eta_j) \qquad (2.3.15)$$

where g_i denotes $\partial g/\partial \eta_i$ evaluated for $\eta_1,\eta_2,\ldots,\eta_k$ equal to their mean values. For this formula to be valid, the quantities $\mathrm{var}\,\eta_i$ should be small in comparison with $\{\mathscr{E}g\}^2$. To obtain it, expand g as a Maclaurin series in the η_i, neglecting the terms of the second degree, and then use (2.3.14).

Important distributions include the *normal distribution*

$$F(y) = \int_{-\infty}^{y} (2\pi\sigma^2)^{-1/2}\exp\{-\tfrac{1}{2}(t-\mu)^2/\sigma^2\}\,dt, \qquad (2.3.16)$$

the *exponential distribution*

$$F(y) = \begin{cases} 0, & y < 0 \\ 1-e^{-\lambda y}, & y \geqslant 0 \end{cases} \qquad (\lambda > 0), \qquad (2.3.17)$$

the *rectangular distribution*

$$F(y) = \begin{cases} 0, y < a \\ (y-a)/(b-a), a \leqslant y \leqslant b, \\ 1, y > b \end{cases} \qquad (2.3.18)$$

and the *binomial distribution*

$$F(y) = \sum_{t \leqslant y} \frac{n!}{t!(n-t)!} p^t (1-p)^{n-t}. \qquad (2.3.19)$$

In (2.3.19), n and t are non-negative integers, and $0 \leqslant p \leqslant 1$; the binomial distribution represents the distribution of the total number η of 'successful' events, when each of n independent random events has a probability p of being 'successful' and a probability $1-p$ of being 'unsuccessful'.

The four distributions given above exhibit a feature common to many distribution functions: the distribution function has some specified mathematical form but depends upon some unspecified constants [μ, σ in (2.3.16); λ in (2.3.17); a, b in (2.3.18); and n, p in (2.3.19)]. These constants are called *parameters* of the distribution.

In a number of cases it is possible, by means of a linear transformation of the random variable, to bring the distribution function into a *standardized form*. Thus the transformation $\zeta = (\eta - \mu)/\sigma$ applied to the random variable η of (2.3.16), yields for ζ a normal distribution

$$F(z) = \int_{-\infty}^{z} (2\pi)^{-1/2} \exp(-\tfrac{1}{2}t^2) \, dt, \qquad (2.3.20)$$

which is the standardized form of the normal distribution. The standardized forms of (2.3.17) and (2.3.18) arise when $\lambda = 1$ and when $a = 0$, $b = 1$ respectively. The standardized rectangular distribution

$$F(y) = \begin{cases} 0, y < 0 \\ y, 0 \leqslant y \leqslant 1 \\ 1, y > 1 \end{cases} \qquad (2.3.21)$$

is particularly important in Monte Carlo work (see Chapter 3).

The so-called *central limit theorem* asserts that (under suitable mild conditions specified in the standard textbooks) the sum of n independent random variables has an approximately normal distribution when n is large; in practical cases, more often than not, $n = 10$ is a reasonably large number, while $n = 25$ is effectively infinite. There are, of course, qualifications to this rule of thumb concerning the size of n. Firstly, the separate random variables comprising the sum should not have too disparate variances: for example, in terms of variance none of them should be comparable with the sum of the rest. Secondly, as $n \to \infty$, the distribution function of the sum tends to normality more rapidly in the region of the mean than in the *tails* of the distribution (i.e. the regions distant from the mean). For vectors, there is a corresponding theorem in which the limiting distribution is the *multinormal distribution*

$$F(\mathbf{y}) = \int_{-\infty}^{\mathbf{y}} |2\pi\mathbf{V}|^{-1/2} \exp\{-\tfrac{1}{2}(\mathbf{t}-\boldsymbol{\mu})'\mathbf{V}^{-1}(\mathbf{t}-\boldsymbol{\mu})\}\, d\mathbf{t},$$

(2.3.22)

where the expression in the exponent is a quadratic form in matrix notation, and $\mathbf{V}$ is the *variance-covariance matrix* of the random variable $\boldsymbol{\eta}$, that is to say the matrix whose (i,j)-element is the covariance between the ith and jth coordinates of $\boldsymbol{\eta}$.

Another important limiting distribution is the *Poisson distribution*

$$F(y) = \sum_{t \leqslant y} e^{-\lambda} \lambda^t / t!,$$

(2.3.23)

which occurs when $n \to \infty$ and $np \to \lambda$ in (2.3.19). In (2.3.23), t is a non-negative integer and λ is any positive constant.

2.4 Estimation

Most Monte Carlo work is concerned with estimating the unknown numerical value of some parameter of some distribution. Viewed in this context, the parameter is called an *estimand*. The available data will consist of a number of observed random variables, constituting the *sample*. The number of observations in the sample is called the *sample size*. The connexion between the sample and the estimand is that the latter is a parameter of the distribution of the random variables constituting the former. For example, the estimand might be the parameter μ of (2.3.16), and the sample (of size n) might consist of independent random variables $\eta_1, \eta_2, \ldots, \eta_n$ each distributed according to (2.3.16). Now μ is the mean of (2.3.16); so it might seem reasonable to estimate μ by the average of the observations

$$\bar{\eta} = (\eta_1 + \eta_2 + \ldots + \eta_n)/n. \qquad (2.4.1)$$

On the other hand, the quantity

$$(w_1\eta_1 + w_2\eta_2 + \ldots + w_n\eta_n)/(w_1 + w_2 + \ldots + w_n) \qquad (2.4.2)$$

is also an average (in fact, a *weighted average*) of the observations, of which (2.4.1) is a special case. The question then arises: can we choose the w_i in some way such that (2.4.2) is a better estimator of μ than (2.4.1); and, indeed, is there some other function $t(\eta_1, \eta_2, \ldots, \eta_n)$ which is even better than (2.4.2)? The answer naturally depends upon what we mean by a 'better' estimator, and we now examine this issue in general.

We can represent the sample by a vector $\boldsymbol{\eta}$ with coordinates $\eta_1, \eta_2, \ldots, \eta_n$; and the estimand θ will be a parameter of the distribution of $\boldsymbol{\eta}$. We call this distribution the *parent distribution* to distinguish it from the sampling distribution (to be defined below). To estimate θ we are going to use some function of the observations, say $t(\boldsymbol{\eta})$. It is important to distinguish two different senses of the function t. We can either regard t as a mathematical function of some unspecified variables $y_1, y_2, \ldots, y_n$ (i.e. $t = t(\mathbf{y})$), in which case we speak of the *estimator* t; or we can think of the numerical value of t when $y_1, y_2, \ldots, y_n$ take the observed values $\eta_1, \eta_2, \ldots, \eta_n$ (i.e. $t = t(\boldsymbol{\eta})$), in which case we speak of the *estimate* t. Our problem is then to find an estimator which provides good estimates of θ, i.e. to choose a function $t(\mathbf{y})$ such that $t(\boldsymbol{\eta})$ is close to θ.

Now since $\boldsymbol{\eta}$ is a random variable, so is $t(\boldsymbol{\eta})$. This statement is to be understood as follows: $\boldsymbol{\eta}$ is the particular observation obtained by some experiment, and is random to the extent that, if we repeated the experiment, we should expect to get a different value of $\boldsymbol{\eta}$. The parent distribution describes how these values of $\boldsymbol{\eta}$ are distributed when we consider (possibly hypothetical) repetitions of the experiment. Since $\boldsymbol{\eta}$ varies from experiment to experiment, so does $t(\boldsymbol{\eta})$;

and consequently $t(\eta)$ has a distribution, called the *sampling distribution*. If $t(\eta)$ is to be close to θ, then the sampling distribution ought to be closely concentrated around θ.

It is not, however, necessary to repeat the experiment in practice,† because we can determine the sampling distribution mathematically in terms of the estimator $t(y)$ and the parent distribution $F(y)$: in fact the sampling distribution is

$$T(u) = P(t(\eta) \leqslant u) = \int\limits_{t(y) \leqslant u} dF(y), \qquad (2.4.3)$$

where the integral is taken over all values of y such that $t(y) \leqslant u$. Thus, given F, we have to find $t(y)$ such that (2.4.3) clusters about θ. The difference between θ and the average value of $t(\eta)$ (average over hypothetically repeated experiments) will be

$$\beta = \mathscr{E}\{t(\eta) - \theta\} = \int \{t(y) - \theta\} dF(y); \qquad (2.4.4)$$

and similarly the dispersion of $t(\eta)$ can be measured by

$$\sigma_t^2 = \operatorname{var}\{t(\eta)\} = \mathscr{E}\{[t(\eta) - \mathscr{E}t(\eta)]^2\} = \mathscr{E}\{[t - \theta - \beta]^2\}$$

$$= \int \{t(y) - \theta - \beta\}^2 dF(y). \qquad (2.4.5)$$

Indeed, $\theta + \beta$ and σ_t^2 are the mean and the variance of the sampling distribution. We call β the *bias* of t, and σ_t^2 the *sampling variance* of t. It is worth noting that (2.4.4) and (2.4.5) are special cases of (2.3.6).

We can now specify what is meant by a good estimator. We say that $t(y)$ is a good estimator if β and σ_t^2 are small; for this means that $T(u)$ will cluster around θ closely. This is not the only way in which we could have accorded merit to an estimator, but it is a simple and convenient way. In particular, if $t(y)$ is such that $\beta = 0$, we speak of an *unbiased estimator*; and, if σ_t^2 is smaller than the sampling variance of any other estimator, we speak of a *minimum-variance estimator*.

† Nevertheless, it is sometimes convenient to repeat the experiment in practice, especially when (2.4.3) is mathematically intractable. Such repetitions can then afford a Monte Carlo estimate of (2.4.3).

SHORT RESUMÉ OF STATISTICAL TERMS 19

According to the above criteria, we prefer to use unbiased minimum-variance estimators. Unfortunately the problem of finding them is a rather difficult problem in the calculus of variations, though it can be solved in certain cases: thus it turns out that (2.4.1) is an unbiased minimum-variance estimator of the mean μ of a normal distribution. Often, however, we have to be content with only a partial solution: for example, instead of considering all possible estimators and picking the one with minimum variance, we may confine our attention to *linear estimators*, i.e. linear functions of the observations $\eta_1, \eta_2, \ldots, \eta_n$, and pick from this restricted set the one with minimum variance. This limited problem of finding *unbiased minimum-variance linear estimators* is much easier to solve. The following special case is sufficiently important to merit explicit mention. Suppose that it is known that the mean of the parent distribution of η_i is $m_i\theta$, where $i = 1, 2, \ldots, n$ and m_i is a known constant (very often $m_i = 1$ in practical applications) and θ is the unknown estimand. Suppose also that the covariance of η_i and η_j is v_{ij}. Write V for the *variance-covariance matrix* of η, i.e. V is the symmetric $n \times n$ matrix whose (i,j)-element is v_{ij}; write **m** for the column-vector $\{m_1, m_2, \ldots, m_n\}$; and **m**' for the transpose of **m**. Then the unbiased minimum-variance linear estimator of θ is

$$t = (\mathbf{m}'\mathbf{V}^{-1}\mathbf{m})^{-1}\mathbf{m}'\mathbf{V}^{-1}\eta, \qquad (2.4.6)$$

and its sampling variance is

$$\sigma_t^2 = (\mathbf{m}'\mathbf{V}^{-1}\mathbf{m})^{-1}. \qquad (2.4.7)$$

The square root of the sampling variance of an estimator is known as the *standard error* of the estimator. It is, of course, the standard deviation of the sampling distribution; but it is convenient to speak of it as standard error, and to reserve the term *standard deviation* for the standard deviation of the parent distribution. The standard error is a useful indicator of how close the estimate is to the unknown estimand. A good working rule, which holds with relatively few exceptions, is that an estimate has only about one chance in twenty of differing from its mean value by more than twice the standard error, and only about one chance in a thousand of more than thrice the standard error. For unbiased estimators, the mean value is equal to the estimand. The figures 1 in 20 and 1 in 1000 are appropriate to the normal distribution: their application to a wide variety of practical situations stems from the central limit theorem, since most estimators are effectively equivalent to a sum of several independent random variables.

The considerations given above extend to the case where we have to estimate several estimands simultaneously from a single set of data. These estimands can be regarded as the co-ordinates of a vector estimand, for which any vector estimator will have a distribution and a sampling variance-covariance matrix, and will be unbiased if each of its co-ordinates is unbiased. The minimum-variance estimator is usually understood to be the one which minimizes the elements on the principal diagonal of this matrix, i.e. the individual sampling variances of individual co-ordinates.

There is one other estimation technique worth mentioning here, namely the *method of maximum likelihood*. The *likelihood* of a sample is a number proportional to the probability of that sample: it can usually be taken to be $f(\eta)$, where f is the function specified in (2.3.8). This number depends upon the observed sample η and also upon the estimand $\theta = \{\theta_1, \theta_2, ..., \theta_p\}$. The maximum-likelihood estimate is the value of θ which maximizes the likelihood for the observed η; and is therefore a function of η. In practice it is often more convenient to maximize L, the natural logarithm of the likelihood; in which case the maximum-likelihood estimator is the solution of the equations

$$\partial L/\partial\theta_i = 0 \quad (i = 1, 2 ..., p). \tag{2.4.8}$$

To a satisfactory approximation, the sampling variance-covariance matrix of this estimator is the reciprocal of the matrix whose (i,j)-element is $-\partial^2 L/\partial\theta_i\partial\theta_j$.

The following particular estimators are of frequent occurrence. Suppose that $\eta_1, \eta_2, ..., \eta_n$ are independent observations from the same parent distribution. Then an unbiased estimator of the mean of this parent distribution is

$$\bar{\eta} = (\eta_1 + \eta_2 + ... + \eta_n)/n \tag{2.4.9}$$

and it has standard error

$$\sigma_{\overline{\eta}} = \sigma/\sqrt{n}, \tag{2.4.10}$$

where σ is the standard deviation of the parent distribution. An unbiased estimator of the variance of the parent distribution is

$$s^2 = (\eta_1^2 + \eta_2^2 + \ldots + \eta_n^2 - n\overline{\eta}^2)/(n-1), \tag{2.4.11}$$

which has standard error approximately equal to

$$\sigma_{s^2} \simeq \sigma^2/\sqrt{(\tfrac{1}{2}n)}. \tag{2.4.12}$$

Whereas the standard error (2.4.10) is exact, whatever the parent distribution, the validity of the approximation (2.4.12) depends upon the fourth moment of the parent distribution. Equation (2.4.12) holds exactly if the parent distribution is normal, and approximately if it is approximately normal. It is usual to take the square root of (2.4.11) as an estimator of σ, even though this is not unbiased† : the standard error of this estimator is approximately

$$\sigma_s \simeq \sigma/\sqrt{(2n)}. \tag{2.4.13}$$

In the binomial distribution (2.3.19), an unbiased estimate of the parameter p is η/n, the ratio of successful trials to the total number of trials; and the standard error of this estimator is

$$\sigma_p = \sqrt{\{p(1-p)/n\}}. \tag{2.4.14}$$

In assigning numerical values to standard errors, one usually replaces the value of a parameter by its estimate: thus in (2.4.10), (2.4.12), and (2.4.13) one would use s, calculated from (2.4.11), in place of σ; and in (2.4.14) one would use η/n in place of p.

2.5 Efficiency

The main concern in Monte Carlo work is to obtain a respectably small standard error in the final result. It is always possible to reduce standard errors by taking the average of n independent values of an estimator; but this is rarely a rewarding procedure, as usually (as in

† If t is an unbiased estimator of θ, it will be generally false that $\phi(t)$ is an unbiased estimator of $\phi(\theta)$; but, if t is a maximum likelihood estimator of θ, then $\phi(t)$ is a maximum likelihood estimator of $\phi(\theta)$. One may therefore expect maximum likelihood estimators to be biased.

22 MONTE CARLO METHODS

(2.4.10), (2.4.12), (2.4.13), and (2.4.14)) the standard error is inversely proportional to the square root of the sample size n and therefore, to reduce the standard error by a factor of k, the sample size needs to be increased k^2-fold. This is impracticable when k is large, say 100. The remedy lies in careful design of the way in which the data is collected and analyzed. The *efficiency* of a Monte Carlo process may be taken as inversely proportional to the product of the sampling variance and the amount of labour expended in obtaining this estimate. It pays handsome dividends to allow some increase in the labour (due, say, to the use of sophisticated rather than crude estimators) if that produces an overwhelming decrease in the sampling variance.

The so-called *variance-reducing techniques*, which lie at the heart of good Monte Carlo work, are techniques which reduce the coefficient of $1/n$ in the sampling variance of the final estimator, where n is the sample size (or, perhaps more generally, the amount of labour expended on the calculation). These techniques depend upon various devices, such as distorting the data so that the variance of the parent distribution is reduced (e.g. in importance sampling), or making allowances for various causes of variation in the data (e.g. as in regression). What most of these methods have in common is that they do not introduce bias into the estimation; and thus they make results more precise without sacrificing reliability.

2.6 Regression

Sometimes the variation in raw experimental data can be broken into two parts: the first part consists of an entirely random variation that we can perhaps do little about; but the second part arises because the observations are influenced by certain concomitant conditions of the experiment, and it may be possible to record these conditions and determine how they influence the raw observations. When this is so, we can then calculate (or at least estimate) this second part and subtract it out from the reckoning, thus leaving only those variations in the observations which are not due to the concomitant conditions.

The basic model for this is to suppose that the random observations η_i ($i = 1, 2, \ldots, n$) are associated with a set of concomitant numbers x_{ij} ($j = 1, 2, \ldots, p$) which describe the experimental conditions under which the observation η_i was taken. It is then assumed that η_i is the sum of a purely random component δ_i with zero expectation and a linear combination $\sum_j \beta_j x_{ij}$ of the concomitant numbers. Here the coefficients β_j are unknown parameters to be estimated from the data itself. Let X denote the $n \times p$ matrix x_{ij}, and let V be the $n \times n$ variance-covariance matrix of the δ_i's. Then the minimum-variance unbiased linear estimator of the vector β is

$$b = (X'V^{-1}X)^{-1}X'V^{-1}\eta \qquad (2.6.1)$$

and its sampling variance-covariance matrix is $(X'V^{-1}X)^{-1}$. In passing, it may be noticed that (2.4.6) is the particular case $X = m$ of (2.6.1). The numbers β are called *regression coefficients*; and, once they have been estimated, we can allow for the concomitant variation.

At first sight it may seem that the assumption that η_i depends linearly upon the x_{ij} is unnecessarily restrictive. But, in fact, this apparent linear dependence is more a matter of notational convenience than anything else: there is no reason why the x_{ij} should not be functionally related – for example, when the x_{ij} are powers of a single number x_i, say $x_{ij} = x_i^j$, one has the special case of *polynomial regression*. Another important special case arises when the data can be classified into strata or categories, and x_{ij} is given the value 1 or 0 according as η_i does or does not belong to the jth category. In this case formula (2.6.1) is known as the *analysis of variance*. In the mixed case, when some but not all x_{ij} are restricted to the values 0 and 1, one has what is called the analysis of covariance.

2.7 Sampling methods

There are two basic types of sampling. In the first, called *fixed* sampling, one lays one's plans for collecting the data and then collects the data accordingly without reference to the actual numerical values of the resulting observations. In the second, called *sequential sampling*, one allows the method of collecting and the amount collected to depend upon the observed numerical values found during collection. For example, with fixed sampling from a binomial distribution one first selects a number n and then carries out that number n of trials and observes how many are successful; but with a particular type of sequential sampling, known as *inverse sampling* from the binomial distribution, one does not fix n but instead carries out trials until a prescribed number are successful, the observation in this case being the total number of trials needed for this purpose.

Another important type of sampling is *stratified sampling*, in which the data are collected to occupy prearranged categories or strata.

CHAPTER 3

Random, Pseudorandom, and Quasirandom Numbers

3.1 General remarks

The essential feature common to all Monte Carlo computations is that at some point we have to substitute for a random variable a corresponding set of actual values, having the statistical properties of the random variable. The values that we substitute are called *random numbers*, on the grounds that they could well have been produced by chance by a suitable random process. In fact, as we shall go on to describe, they are *not* usually produced in this way; however, this should not affect the person who has to use them, since the question he should be asking is not 'Where did these numbers come from?' but 'Are these numbers correctly distributed?', and this question is answered by statistical tests on the numbers themselves. But even this approach runs into insuperable practical difficulties because strictly speaking it requires us to produce infinitely many random numbers and make infinitely many statistical tests on them to ensure fully that they meet the postulates. Instead we proceed with a mixture of optimism and utilitarianism: optimism, in the sense that we produce only finitely many numbers, subject them to only a few tests, and hope (with some justification) that they would have satisfied the remaining unmade tests; utilitarianism, in the sense that *one* of the tests that might have been applied is whether or not the random numbers yield an unbiased or a reliable answer to the Monte Carlo problem under study, and it is really only this test that interests us when we are ultimately concerned only with a final numerical solution to a particular problem. Taken in this second vein, the other tests are irrelevant; the numbers produced need not satisfy them. The more of the irrelevant tests we ignore, or indeed deliberately permit to be violated, the easier and the cheaper it may become to generate the numbers, which now of course are no longer genuinely random but only 'pseudorandom' or 'quasirandom'. It is simply a question of sacrificing ideals to expediency. In Monte Carlo work justice will have been done if the final verdict is fair, i.e. the answers come out approximately right.

When we use the term 'random number' without qualification, we shall be referring to the standardized rectangular distribution (2.3.21). We shall reserve the symbol ξ, with or without suffices, for such a random number. We shall also use the same symbol for some substitute number (such as a pseudorandom number), which in practice plays the role of a standardized rectangular random number. For instance, it is convenient to talk as if numbers were continuously distributed when in fact they can only take on values that are multiples of some small number (say 10^{-10}, or even 10^{-3}), on account of the rounding off necessary for numerical computation. This does not lead to any confusion.

For Monte Carlo work with pencil and paper there are published tables of random numbers, the best known being [1] and [2]. The Rand tables are also available on punched cards, and should be regarded as standard stock in a computing laboratory which does Monte Carlo computations on punched-card machinery. These tables are generated by physical processes which are, as far as one can tell, random in the strict sense, but they have also been successfully subjected to a number of statistical tests.

It is possible to generate one's own random numbers by a like process (and indeed several physical devices† have been constructed for just this purpose), but one then has the additional task of repeatedly verifying that the process is functioning properly, and the practice is not recommended for serious work. (See Lord Kelvin's remarks [3] on the difficulties of picking pieces of paper at random from a jar.)

For electronic digital computers it is most convenient to calculate a sequence of numbers one at a time as required, by a completely specified rule which is, however, so devised that no reasonable statistical test will detect any significant departure from randomness.

† See the section on 'Physical devices for generating random numbers' on p. 159 in the Further References.

Such a sequence is called *pseudorandom*. The great advantage of a specified rule is that the sequence can be exactly reproduced for purposes of computational checking.

Ordinarily, and in the following section §3.2, we compute a pseudorandom sequence ξ_i from a sequence of positive integers x_i via the relation

$$\xi_i = x_i/m, \tag{3.1.1}$$

where m is a suitable positive integer.

Pseudorandom sequences are intended for general use on all classes of problem. However, in some kinds of Monte Carlo work, where we know that the violation of some statistical tests will not invalidate the result, it may be to our advantage deliberately to use a non-random sequence having only the particular statistical properties that concern us. Such a sequence is called *quasirandom*.

Hull and Dobell [4] give an extensive bibliography on random number generators.

3.2 Pseudorandom numbers

The first suggestion for producing a pseudorandom sequence, due to Metropolis and von Neumann, was the 'midsquare' method (see, for instance, Forsythe [5]), in which each number is generated by squaring its predecessor and taking the middle digits of the result: that is to say, x_{i+1} in (3.1.1) consists of the middle digits of x_i^2. (All pseudorandom number generators exploit a fixed word-length in a digital computer.) This method was soon found to be unsatisfactory, and later work, reviewed in [4], has been concentrated on congruential methods, to be described in more detail below. Apart from these, various *ad hoc* methods have been devised, making use of the peculiarities of individual computers; these are generally lacking in theoretical support, but there is no objection to using them if they satisfy reasonable statistical requirements.

In 1951, Lehmer [6] suggested that a pseudorandom sequence could be generated by the recurrence relation†

$$x_i \equiv ax_{i-1} \text{ (modulo } m), \tag{3.2.1}$$

† The notation signifies that x_i is the remainder when ax_{i-1} is divided by m.

and this has subsequently been generalized [7] to

$$x_i \equiv ax_{i-1} + c \pmod{m}. \tag{3.2.2}$$

Here m is a large integer determined by the design of the computer (usually a large power of 2 or of 10) and a, c, and x_i are integers between 0 and $m-1$. The numbers x_i/m are then used as the pseudo-random sequence. Formulae (3.2.1) and (3.2.2) are called *congruential* methods of generating pseudorandom numbers; in particular (3.2.1) is the *multiplicative congruential method*.

Clearly, such a sequence will repeat itself after at most m steps, and will therefore be periodic; for example, if $m = 16$, $a = 3$, $c = 1$, and $x_0 = 2$, the sequence of x's generated by (3.2.2) is 2, 7, 6, 3, 10, 15, 14, 11, 2, 7, ..., so that the period is 8. We must always ensure that the period is longer than the number of random numbers required in any single experiment. The value of m is usually large enough to permit this.

If recurrence (3.2.2) is used, the full period of m can always be achieved, provided that:

$$\left.\begin{array}{l} \text{(i)} \quad c \text{ and } m \text{ have no common divisor;} \\ \text{(ii)} \quad a \equiv 1 \pmod{p} \text{ for every prime factor } p \text{ of } m; \\ \text{(iii)} \quad a \equiv 1 \pmod{4} \text{ if } m \text{ is a multiple of 4.} \end{array}\right\} \tag{3.2.3}$$

The case of (3.2.1) is rather more complicated. The period is now always less than m, and if

$$m = 2^\alpha p_1^{\beta_1} \ldots p_r^{\beta_r}, \tag{3.2.4}$$

where the p's are distinct odd primes, the maximum possible period is

$$\lambda(m) = \text{lowest common multiple of } \lambda(2^\alpha), \lambda(p_1^{\beta_1}), \ldots, \lambda(p_r^{\beta_r}), \tag{3.2.5}$$

where
$$\left.\begin{array}{l} \lambda(p^\beta) = p^{\beta-1}(p-1) \text{ if } p \text{ is odd,} \\ \lambda(2^\alpha) = 1(\alpha = 0, 1), \\ \qquad\quad = 2(\alpha = 2), \\ \qquad\quad = 2^{\alpha-2}(\alpha > 2). \end{array}\right\} \tag{3.2.6}$$

This maximum period is achieved provided that:

(i) $a^n \not\equiv 1$ (modulo $p_j^{\beta_j}$) for $0 < n < \lambda(p_j^{\beta_j})$;

(ii) $a \equiv 1$ (modulo 2) if $\alpha = 1$,
$\equiv 3$ (modulo 4) if $\alpha = 2$,
$\equiv 3$ or 5 (modulo 8) if $\alpha > 2$;

(iii) x_0 is prime relative to m.

$$(3.2.7)$$

We refer the reader to [4] for proofs of these statements.

The case of most frequent occurrence is $m = 2^\alpha$, since the requisite arithmetic can then be performed very easily on a computer working in the binary scale by means of a command sometimes called *low multiplication*, α being generally in the neighbourhood of 30 or 40. Then (3.2.1) gives a period of $m/4$ provided that a differs by 3 from the nearest multiple of 8 and that x_0 is odd, and (3.2.2) gives a period of m provided that c is odd and that a is one greater than a multiple of 4.

The condition that the period be a maximum ensures that the numbers produced are rectangularly distributed (to within the accuracy of the machine), but they may nevertheless be highly correlated, and much work remains to be done on this question. Greenberger [8] has derived an approximate result, that the correlation between successive numbers generated by (3.2.2) lies between the bounds

$$\frac{1}{a} - \frac{6c}{am}\left(1 - \frac{c}{m}\right) \pm \frac{a}{m}, \tag{3.2.8}$$

from which one can also find approximate bounds for the correlation between numbers k steps apart by replacing a and c by a^k and $(a^k - 1)c/(a - 1)$, both reduced modulo m. However, the result is of more use in rejecting bad choices of c and a than in finding good choices.

We are thus driven back to carry out statistical tests on the actual numbers produced. We shall not go into details of all the possible tests. A fairly extensive account of these and a bibliography is given by Taussky and Todd [9], and here we may content ourselves with a simple illustration.

Let us investigate the possibilities of gross irregularities of distribution and serial correlation between numbers generated by (3.2.1) with $m = 2^{29}$ and $a = 129140163$ ($= 3^{17}$). (We choose (3.2.1) rather than (3.2.2) only because it happens to be more convenient for the computer we have to hand.) Accordingly, we take the first 10,000 numbers generated by this process and count the number m_j which satisfy

$$(j-1)/25 \leqslant \xi_i < j/25 \quad (j = 1, 2, \ldots, 25) \qquad (3.2.9)$$

and the number n_{jk} which satisfy

$$(j-1)/5 \leqslant \xi_i < j/5 \quad \text{and} \quad (k-1)/5 \leqslant \xi_{i+1} < k/5$$

$$(j, k = 1, 2, \ldots, 5). \qquad (3.2.10)$$

This gives us two sets of 25 numbers, shown in Tables 3.1 and 3.2.

Table 3.1

392	423	386	396	425
386	400	393	416	363
411	389	385	363	441
437	387	385	399	416
396	406	405	415	385

$$\chi^2 = 23\cdot4$$

24 d.f.:   95% level 13·85
5% level 36·42

We then proceed to compute the statistic χ^2 for each table. This is defined for Table 3.1 by

$$\chi^2 = \sum_j (m_j - \bar{m}_j)^2 / \bar{m}_j, \qquad (3.2.11)$$

where $\bar{m}_j$ is the expected value of m_j, on the assumption that the ξ's are independently and rectangularly distributed. In this case $\bar{m}_j = 400$.

Table 3.1 has 24 degrees of freedom, since the only automatically satisfied condition is that the sum of all the entries is 10000. We therefore refer to tables of the χ^2 distribution for 24 degrees of freedom to find bounds between which χ^2 ought to lie with high probability.

Analysis of Table 3.2 is not so simple and mistaken conclusions have sometimes been drawn from too hasty assumptions. We refer

Table 3.2

	$k = 1$	2	3	4	5	sum
$j = 1$	400	415	408	398	401	2022
2	418	391	383	385	381	1958
3	404	347	408	415	415	1989
4	396	386	402	429	411	2024
5	404	418	388	398	399	2007

$$\chi^2 = 16\cdot4 - 1\cdot5 = 14\cdot9$$
20 d.f.: 95% level 10·85
5% level 31·41

the reader to Good [10] for justification of the correct test, that if $n_j = \sum_k n_{jk}$ then, in the present case, the statistics

$$\sum_j \sum_k (n_{jk} - 400)^2/400 - \sum_j (n_j - 2000)^2/2000 \qquad (3.2.12)$$

and

$$\sum_j \sum_k (n_{jk} - 400)^2/400 - 2\sum_j (n_j - 2000)^2/2000 \qquad (3.2.13)$$

have approximate χ^2 distributions with 20 and 16 degrees of freedom respectively. We have chosen to use (3.2.12).

Both of our values of χ^2 lie well within the bounds indicated, so that the generator passes these tests.

It is well to observe that, while it is possible to construct a sequence of random numbers out of a sequence of random digits in the obvious manner, it is not necessarily true that the successive digits of a sequence of pseudorandom numbers are random. For example, the successive binary digits of numbers generated by (3.2.1) with $m = 2^\alpha$ have shorter and shorter periods until the last digit is always 1. Thus the last 15 or 20 digits of each random number should be discarded when constructing sequences of random digits.

3.3 Quasirandom numbers

Consider the estimation of

$$\theta = \int_0^1 \ldots \int_0^1 f(\mathbf{x})\, dx_1 \ldots dx_k \qquad (3.3.1)$$

by a formula

$$\frac{1}{N}\{f(\xi_1) + \ldots + f(\xi_N)\} \tag{3.3.2}$$

where $\mathbf{x}$ and ξ_n are k-dimensional vectors. If the co-ordinates of the ξ_n are kN terms of a random sequence, we can use statistical arguments to show that the error will be of the order of $N^{-1/2}$. In Chapter 5 we shall show how this error might be reduced by the use of antithetic variates, when there is functional dependence between several successive ξ's. However, except for small values of k, present antithetic techniques require N to be hopelessly large before they can be applied at all.

Suppose, however, that we abandon the idea of a general-purpose random sequence, and concentrate on sequences which estimate (3.3.1) as well as possible for all values of N. In what follows we use the vector inequality $\mathbf{x} \leqslant \mathbf{y}$ to signify that no co-ordinate of $\mathbf{x}$ exceeds the corresponding co-ordinate of $\mathbf{y}$. In order to have a manageable criterion, let $f(\mathbf{x})$ be restricted to functions of the class

$$\left. \begin{array}{l} f(\mathbf{x}) = 1 \text{ if } \mathbf{x} \leqslant \mathbf{A} \\ \phantom{f(\mathbf{x})} = 0 \text{ otherwise} \end{array} \right\}. \tag{3.3.3}$$

The condition that (3.3.2) shall give a good estimate of the integral of every function of this class can be expressed in other words as that the points $\xi_1, \ldots, \xi_N$ are evenly distributed over the hypercube.

Let $S(\mathbf{A})$ be the number of the first N points of the sequence $\{\xi_n\}$ satisfying $\xi_n \leqslant \mathbf{A}$. Then the integral (3.3.1) is $A_1 A_2 \ldots A_k$, the product of the co-ordinates of $\mathbf{A}$, while the estimator (3.3.2) is $S(\mathbf{A})/N$. Let M denote the maximum of

$$|S(\mathbf{A}) - NA_1 A_2 \ldots A_k| \tag{3.3.4}$$

taken over all points $\mathbf{A}$ of the hypercube $\mathbf{0} \leqslant \mathbf{A} \leqslant \mathbf{1}$, and let

$$J = \int_0^1 \ldots \int_0^1 [S(\mathbf{A}) - NA_1 A_2 \ldots A_k]^2 \, d\mathbf{A} \tag{3.3.5}$$

denote the mean square of (3.3.4) over the hypercube. Both M and J depend upon the particular sequence $\{\xi_n\}$ used, and can serve as criteria of the effectiveness of the sequence. The smaller these values, the better is the sequence for integrating functions of the class (3.3.3). Since other types of integrands can be approximated by linear combinations of functions similar to (3.3.3), the criteria will be fairly widely applicable for largish values of N.

Roth [11] has shown that there are constants c_k such that

$$J > c_k (\log N)^{k-1} \qquad (3.3.6)$$

for any sequence of ξ's whatever. It is not known whether this bound can be achieved.

The best that has been done towards minimizing M and J is based on an idea of van der Corput [12]. Suppose that the natural numbers are expressed in the scale of notation with radix R, so that

$$n = a_0 + a_1 R + a_2 R^2 + \ldots + a_m R^m \quad (0 \leqslant a_i < R). \quad (3.3.7)$$

Now write the digits of these numbers in reverse order, preceded by a point. This gives the numbers

$$\phi_R(n) = a_0 R^{-1} + a_1 R^{-2} + \ldots + a_m R^{-m-1}. \qquad (3.3.8)$$

For example, in the binary scale ($R = 2$),

n = 1 (decimal) =	1 (binary);	$\phi_2(n)$ = 0·1 (binary) =	0·5 (decimal)
2	10	0·01	0·25
3	11	0·11	0·75
4	100	0·001	0·125
5	101	0·101	0·625
6	110	0·011	0·375
7	111	0·111	0·875
8	1000	0·0001	0·0625

and so on.

Halton [13], extending to k dimensions the results found by van der Corput in 2 dimensions, has shown that, provided that $R_1, R_2, \ldots, R_k$ are mutually coprime, the sequence of vectors

$$(\phi_{R_1}(n), \phi_{R_2}(n), \ldots, \phi_{R_k}(n)) \quad (n = 1 \text{ to } N) \qquad (3.3.9)$$

has
$$M < (\log N)^k \prod_{i=1}^{k} \left(\frac{3R_i - 2}{\log R_i} \right)$$

$$\tag{3.3.10}$$

and
$$J < (\log N)^{2k} \prod_{i=1}^{k} \left\{ \frac{(R_i - 1)^2}{\log R_i} \right\}$$

while the sequence

$$(n/N, \phi_{R_1}(n), \ldots \phi_{R_{k-1}}(n)) \tag{3.3.11}$$

satisfies the same inequalities (3.3.10) with k replaced by $(k-1)$.

For comparison, it is easily shown that, in the case of a random sequence, the expectation of J is the much larger quantity

$$N(2^{-k} - 3^{-k}). \tag{3.3.12}$$

A different approach is based on the theory of Diophantine approximation. Richtmyer [14] considers the sequence of vectors

$$\boldsymbol{\xi}_n = ([n\alpha_1], [n\alpha_2], \ldots, [n\alpha_k]), \tag{3.3.13}$$

where $[\cdot]$ denotes the fractional part of the enclosed number and $\alpha_1, \alpha_2, \ldots, \alpha_k$ are independent irrational numbers belonging to a real algebraic field of degree $\delta (\geqslant k+1)$. (That is to say, $\alpha_1, \ldots, \alpha_k$ are real roots of a polynomial of degree δ with integer coefficients, and the equation

$$\mathbf{n} \cdot \boldsymbol{\alpha} = n_1 \alpha_1 + n_2 \alpha_2 + \ldots + n_k \alpha_k = 0 \tag{3.3.14}$$

has no solution in integers, other than $n_1 = n_2 = \ldots = n_k = 0$.) Suppose that $f(\mathbf{x})$ has the Fourier expansion

$$f(\mathbf{x}) = \sum_{n_1} \ldots \sum_{n_k} a(n_1, \ldots, n_k) e^{2\pi i n \cdot \mathbf{x}}. \tag{3.3.15}$$

It can then be shown that if

$$\sum_{n_1} \ldots \sum_{n_k} (\max_j |n_j|^{\delta - 1}) |a(n_1, \ldots, n_k)| < \infty \tag{3.3.16}$$

then the difference between (3.3.1) and (3.3.2) is $O(N^{-1})$ as $N \to \infty$. More usefully, if the series (3.3.15) is absolutely convergent, so that

$$\sum_{n_1} \ldots \sum_{n_k} |a(n_1, \ldots, n_k)| = B < \infty, \tag{3.3.17}$$

then (3.3.2) is an unbiased estimator of (3.3.1) and the expected error is less than

$$(1 + \log N)\, B/N. \qquad (3.3.18)$$

Comparing this method and crude Monte Carlo (see Chapter 5) with numerical quadrature formulae using the same number of points, Richtmyer claims that his method is more accurate (for large N) than the trapezoidal rule in two or more dimensions, and than second-order rules in three or more dimensions, while crude Monte Carlo becomes better than the trapezoidal rule in three dimensions and better than second-order rules in five dimensions.

Irrational numbers cannot, of course, be represented exactly in a computer or on paper, but the rounding error can be neglected provided that the number of significant digits exceeds the number of digits in N^2.

Going further along the same lines, Haselgrove [15] considers estimators of θ of the form

$$s(N) = \sum_m c_{Nm} f([\tfrac{1}{2}m\alpha_1 + \tfrac{1}{2}], \ldots, [\tfrac{1}{2}m\alpha_k + \tfrac{1}{2}]), \qquad (3.3.19)$$

of which the simplest instances are

$$s_1(N) = (2N+1)^{-1} \sum_{m=-N}^{N} f([\tfrac{1}{2}m\alpha_1 + \tfrac{1}{2}], \ldots) \qquad (3.3.20)$$

and

$$s_2(N) = (N+1)^{-2} \sum_{m=-N}^{N} (N+1-|m|) f([\tfrac{1}{2}m\alpha_1 + \tfrac{1}{2}], \ldots). \qquad (3.3.21)$$

Here again $\alpha_1, \ldots, \alpha_k$ are linearly independent irrational numbers, but not necessarily algebraic. The first of these estimators, $s_1(N)$, is essentially the same as Richtmyer's.

Haselgrove shows that if the Fourier coefficients satisfy

$$|a(n_1, \ldots, n_k)| \leqslant M_t |n_1 \ldots n_k|^{-t} \qquad (3.3.22)$$

(ignoring zero factors in the product on the right-hand side), and if $s(N)$ is of 'order' $r < t$, defined by

$$\sum c_{Nm} e^{im\phi} = O\{(N \sin \tfrac{1}{2}\phi)^{-r}\} \text{ as } N \to \infty, \qquad (3.3.23)$$

then $|s(N) - \theta| = O(N^{-r})$. In particular, the orders of $s_1(N)$ and $s_2(N)$

are respectively 1 and 2, so that for suitable functions f they give estimates of θ whose errors are $O(N^{-1})$ and $O(N^{-2})$.

He further shows that, for all functions f satisfying (3.3.22) with $t = 2$, $|s_2(N) - \theta|$ is greatest when $f(\mathbf{x})$ is proportional to

$$f_2(\mathbf{x}) = \prod_{i=1}^{k} \{(1 - |2x_i - 1|)^2\}. \qquad (3.3.24)$$

Consequently, if one can find a vector $\boldsymbol{\alpha}$ that minimizes $\sup_N |s_2(N) - \theta|$ for the case $f = f_2$, then this $\boldsymbol{\alpha}$ will lead to a good estimator for the integral of all sufficiently well-behaved functions. A table of suitable values of $\boldsymbol{\alpha}$, determined empirically for $k = 1$ to 8, is given in [15].

3.4 Sampling from non-rectangular distributions

We have so far concentrated on rectangularly distributed random numbers. Frequently we require to take samples from other specified distributions. This is always done by taking rectangularly distributed numbers and transforming them somehow.

We continue writing ξ (with or without suffix) for a rectangularly distributed random number, and we let η denote a number with frequency function $f(y)$ and distribution $F(y)$. The problem is then to express η as an explicit function of ξ's.

If F has a known inverse function F^{-1}, this is easy; we simply take $\eta = F^{-1}(\xi)$. How satisfactory this is, depends upon how easy it is to compute $F^{-1}(\cdot)$. It is worth noting that ξ and $1 - \xi$ have the same distribution; so that we may replace $F^{-1}(\xi)$ by $F^{-1}(1 - \xi)$ if we wish. This device is useful in sampling by computer from the exponential distribution (2.3.17), which describes the distance travelled by a neutron between successive collisions in a nuclear reactor. We have

$$\eta = \lambda^{-1} \log \xi, \qquad (3.4.1)$$

and we can employ a standard subroutine to evaluate the logarithm in (3.4.1).

If f is bounded and η has a finite range, we may use the *rejection technique* of von Neumann [16]. Let a be such that $af(y) \leqslant 1$, and let the range of η be (c, d). We successively choose independent random numbers ξ_1, ξ_2, until we first encounter a pair (ξ_{2r-1}, ξ_{2r}) satisfying

$$\xi_{2r-1} \leqslant af[c + (d - c)\xi_{2r}]. \qquad (3.4.2)$$

We then take $\eta = c + (d-c)\xi_{2r}$, all previous pairs being rejected. This technique tends to be very extravagant if f has a wide variation; however, the principle of the method is implicit in some of the others mentioned below.

One of the earliest rejection techniques, due to von Neumann [16], generates the sine and cosine of a uniformly distributed angle θ. We take

$$\cos\theta = \frac{\xi_1^2 - \xi_2^2}{\xi_1^2 + \xi_2^2}, \qquad \sin\theta = \frac{\pm 2\xi_1\xi_2}{\xi_1^2 + \xi_2^2} \qquad (3.4.3)$$

whenever the independent pair ξ_1, ξ_2 satisfy $\xi_1^2 + \xi_2^2 \leqslant 1$, any pair such that $\xi_1^2 + \xi_2^2 > 1$ being rejected. The sign of $\sin\theta$ is chosen at random.

Butler [17] has described the *composition method*. Suppose that

$$f(y) = \int g_z(y)\,dH(z),$$

where $\{g_z(y)\}$ is a family of density functions and $H(z)$ is a distribution function, from all of which we know how to sample. Then we may obtain η in two stages, first sampling ζ from the distribution H and then sampling η from the corresponding distribution g_ζ.

A special case of this, in which H is a discrete distribution, occurs when $f(y)$ is restricted to a range which we may suppose to be the interval $(0,1)$, and has a power-series expansion

$$f(y) = \sum_{n=0}^{\infty} a_n y^n, \qquad (3.4.4)$$

all of whose coefficients a_n are positive. In this case we may choose n from a discrete distribution with frequency function

$$p_n = (n+1)^{-1} a_n / \sum_i (i+1)^{-1} a_i \qquad (n \geqslant 0) \qquad (3.4.5)$$

and then sample η from the distribution

$$g_n(y) = (n+1) y^n. \qquad (3.4.6)$$

The latter is most simply (though rather uneconomically) performed by taking $\eta = \max(\xi_1, \ldots, \xi_{n+1})$, which may easily be shown to have the density (3.4.6). For example, the distribution

$$f(x) = e^x/(e-1) = \frac{1}{e-1}\{1+x+x^2/2!+\ldots\} \qquad (0 \leqslant x < 1)$$

$$(3.4.7)$$

may be generated in this way, and hence, by the transformation $x'_m = m - x$, the distribution with density

$$g_m(x) = e^{m-x}/(e-1) \quad (m-1 < x \leqslant m). \qquad (3.4.8)$$

We may now perform a second composition, sampling m from the distribution

$$p_m = (e-1)e^{-m} \quad (m = 1, 2, \ldots), \qquad (3.4.9)$$

so that the final result has density function

$$f(y) = \sum_{m=1}^{\infty} g_m(y)p_m = e^{-y} \quad (0 < y \leqslant \infty), \qquad (3.4.10)$$

the exponential distribution. Putting this more concisely [18], the number

$$\eta = m - \max(\xi_1, \ldots, \xi_n) \qquad (3.4.11)$$

is exponentially distributed when m and n are distributed independently according to the laws with frequency functions p_m and q_n given by

$$p_m = (e-1)e^{-m} \quad (m = 1, 2, \ldots)$$

and

$$(3.4.12)$$

$$q_n = \frac{1}{(e-1)\,n!} \quad (n = 1, 2, \ldots).$$

More generally, to sample from a particular discrete distribution with any prescribed frequency function $p_n(n = 1, 2, \ldots)$, we may keep a record of the quantities

$$P(n) = \sum_{i=1}^{n} p_i \qquad (3.4.13)$$

Having calculated a random number ξ, we then determine the required random n as the integer satisfying

$$P(n-1) < \xi \leqslant P(n) \qquad (3.4.14)$$

where $P(0) = 0$. This method is quite fast if $P(n)$ is nearly 1 for some small value of n. For example, the first distribution in (3.4.12) gives $P(1) = 0.63$, $P(2) = 0.86$, $P(3) = 0.95$, $P(4) = 0.98$, ...; while the second distribution of (3.4.12) gives $P(1) = 0.58$, $P(2) = 0.87$, $P(3) = 0.97$, $P(4) = 0.99$, ...

Composition methods can be particularly quick and efficient on a computer with a random access store large enough to hold fairly elaborate look-up tables. Marsaglia [19] has made considerable study of composition methods of this type where H is discrete, his object being to express

$$f(y) = \sum g_n(y) p_n \tag{3.4.15}$$

in such a way that $\sum T_n p_n$ is small, where T_n is the expectation of the computational time taken in sampling from g_n.

For distributions with a finite range, Teichroew [20] has a general method, in which he finds an elementary combination ζ of ξ's, whose distribution is fairly close to F, and then improves this by taking η to be the polynomial in ζ giving a distribution which is a best fit to F in the sense of Chebyshev.

Of special importance is the generation of samples from the *normal distribution* (2.3.20). Muller [21] gives an account of several methods of generating random normal deviates, of which the most attractive is the method of Box and Muller [22], who take

$$\left.\begin{array}{l} \eta_1 = (-2 \log \xi_1)^{1/2} \cos(2\pi \xi_2) \\ \eta_2 = (-2 \log \xi_1)^{1/2} \sin(2\pi \xi_2) \end{array}\right\}. \tag{3.4.16}$$

This method produces normal deviates in independent pairs, η_1, η_2, but is not as fast as Marsaglia's method [19] based on (3.4.15). Another method, which is very simple to program and is not unduly slow, relies on the central limit theorem (see §2.3): we merely take

$$\eta = \xi_1 + \xi_2 + \ldots + \xi_n - \tfrac{1}{2}n. \tag{3.4.17}$$

Here $n = 12$ is a reasonably large number for the purposes of the central limit theorem, and has the advantage that with this choice of n the variance of n is 1, as required by (2.3.20), since the variance of a standardized rectangular random number is $1/12$. The term $-\tfrac{1}{2}n$ in

(3.4.17) ensures that $\mathscr{E}\eta = 0$. Of course, (3.4.17) is inappropriate if the Monte Carlo problem critically depends in some way upon the extreme tails of the normal distribution.

Butcher [23] has given another method which combines the composition and rejection techniques. If

$$f(y) = \sum \alpha_n f_n(y) g_n(y), \tag{3.4.18}$$

where $\alpha_n > 0$, f_n is a density function, and $0 \leqslant g_n \leqslant 1$, then one may choose n with probability

$$p(n) = \alpha_n / \sum \alpha_n, \tag{3.4.19}$$

and choose η from a distribution with density f_n. One now takes a random number ξ, and if $\xi > g_n(y)$ one rejects η and repeats the whole process. To obtain the normal distribution, one may take

$$\left.\begin{array}{ll}
& \alpha_1 = \mu\sqrt{(2/\pi)} \\
& f_1(y) = 1/2\mu \quad (-\mu < y < \mu) \\
& g_1(y) = \exp(-y^2/2), \\
\text{and} & \alpha_2 = \lambda^{-1}\sqrt{(2/\pi)}\exp(\tfrac{1}{2}\lambda^2 - \lambda\mu) \\
& f_2(y) = \tfrac{1}{2}\lambda\exp\{-\lambda(|y|-\mu)\} \quad (y > \mu \text{ or } y < -\mu) \\
& g_2(y) = \exp\{-\tfrac{1}{2}(|y|-\lambda)^2\},
\end{array}\right\} \tag{3.4.20}$$

where λ and μ are arbitrary parameters. The particular choice $\lambda = \sqrt{2}$, $\mu = 1/\sqrt{2}$ minimizes the number of rejections.

Which is the best of several devices for sampling from a given distribution depends very much on the type of computer available and on the proportions of the Monte Carlo work which have to be devoted to such sampling on the one hand and to the manipulation of the sample on the other hand. The choice must rest on the context of the problem, and any general recommendations given here might well prove misleading. With pencil-and-paper Monte Carlo work the situation is simpler; if they are available, one simply uses tables of random numbers having the specified distribution. For the normal distribution, [2] and [24] are suitable.

3.5 Equidistribution of pseudorandom numbers

Let $\varXi = \{\xi_i\}$ $(i = 1, 2, \ldots)$ be an infinite sequence of numbers. Let I denote an arbitrary subinterval of the unit k-dimensional hypercube, and define

$$\chi_i(I, \varXi) = \begin{cases} 1 \text{ if } (\xi_i, \xi_{i+1}, \ldots, \xi_{i+k-1}) \in I \\ 0 \text{ otherwise.} \end{cases} \tag{3.5.1}$$

The sequence $\varXi$ is said to be *equidistributed by* k's if

$$\lim_{n \to \infty} n^{-1} \sum_{i=1}^{n} \chi_i(I, \varXi) = |I| \tag{3.5.2}$$

for every subinterval I, where $|I|$ denotes the volume of I. A sequence which is equidistributed by k's for every k is called *completely equidistributed*.

Completely equidistributed sequences satisfy all the standard statistical tests in the limit for infinitely large samples, although it is possible to devise special tests of randomness not satisfied by any given such sequence. With the reasonable assumption that failure in such a test will not matter except in most extraordinary circumstances, it may seem sufficient to use any completely equidistributed sequence as a source of random numbers. Franklin [25] gives a survey of the interesting theoretical work that has been done on equidistribution and related kinds of uniformity, contributing a number of new results of his own. For instance, he shows that

$$\xi_i = \{\theta^i\} \quad (i = 1, 2, \ldots), \tag{3.5.3}$$

where $\{x\}$ denotes the fractional part of x, is completely equidistributed for almost all $\theta > 1$, although it will not be equidistributed by k's if θ is an algebraic number of degree less than k.

Such theoretical results are, however, not quite applicable to practical work, since the theory covers the asymptotic properties of infinite sequences only; in practice one can dispense with asymptotic properties but is vitally concerned that the finite portion of the sequence that is actually used should behave sufficiently randomly, but the theory tells one nothing about finite sequences. Accordingly, the kind of numerical tests discussed in § 3.2 or the kind of theory in §§3.3 and 3.4 seem more relevant to the situation.

Furthermore, although (3.5.3) leads to an equidistributed sequence for almost all $\theta > 1$, there is (to the best of our knowledge) no *known* value of θ for which (3.5.3) is equidistributed. Until at least one such specific value is found, there is no hope of applying (3.5.3) to a practical situation. The position is very like that for normal numbers; it is known that almost all real numbers are *absolutely normal* (i.e. normal in every scale of notation simultaneously), and there are methods [26], [27] for constructing normal numbers in any given scale, but, again to the best of our knowledge, there is no known example of an absolutely normal number. Here there are opportunities for fascinating theoretical researches, although they may have no immediate practical utility.

3.6 Important concluding principle

We shall later encounter weighting functions and importance sampling, whose purpose is to allow us to sample from one distribution when we 'ought' to be sampling from another. A general Monte Carlo tenet is: *never sample from a distribution merely because it arises in the physical context of a problem, for we may be able to use a better distribution in the computations and still get the right answer.*

CHAPTER 4

Direct Simulation

4.1 General remarks

As explained in §1.1, direct simulation of a probabilistic problem is the simplest form of the Monte Carlo method. It possesses little theoretical interest and so will not occupy us much in this book, but it remains one of the principal forms of Monte Carlo practice because it arises so often in various operational research problems, characterized by a fairly simple general structure overlaid with a mass of small and rather particular details. These problems are beyond the reach of general theory on account of the details, but often are easily simulated and such improvements as might be made by more sophisticated Monte Carlo refinements are rarely worth the effort of their devising. We shall give a few typical examples below. We also make one or two remarks on the rather different use of direct simulation to fortify or quantify qualitative mathematical theories.

4.2 Miscellaneous examples of direct simulation

In [1] there is an account of Monte Carlo calculations applied to the control of floodwater and the construction of dams on the Nile. This is a probabilistic problem because the quantity of water in the river varies randomly from season to season. The data consisted of records of weather, rainfall, and water levels extending over 48 years. The problem is to see what will happen to the water if certain dams are built and certain possible policies of water control exercised; there are a large number of combinations of dams and policies to be examined, and each has to be examined over a variety of meteorological conditions to see what happens not only in a typical year but also in extremely dry or extremely wet years. Finally the behaviour of each system has to be assessed in terms of engineering costs and agricultural,

43

hydroelectric, and other economic benefits. This demands a large quantity of straightforward calculation on a high-speed computer using direct Monte Carlo simulation. The many practical details of the different dams, the course and character of the river bed, losses by evaporation, and so on, preclude the use of any theoretical mathematical model.

Similar Monte Carlo simulation arises in the analysis of storage systems and inventory policy [2] and in the study of bottlenecks and queueing systems in industrial production processes. To deal with the latter, Tocher [3], [4] has written a general computer program, in which various parameters may be set to simulate particular queueing situations.

The study of ecological competition between species or of the progress of epidemics in a community raises very difficult mathematical problems, which may however be resolved quite easily in particular cases by simulation. Bartlett [5] gives further details and examples.

Direct simulation plays a large part in war-gaming [6] and in other forms of operational research. Malcolm [7] gives a bibliography.

Most examples of direct simulation cannot adequately be discussed without going into the many small practical details that necessitated the simulation. We choose, however, one example [8] on the lifetime of comets where this is not so. A long-period comet describes a sequence of elliptic orbits with the sun at one focus. The energy of the comet is inversely proportional to the length of the semimajor axis of the ellipse. For most of the time, the comet is moving at a great distance from the sun; but for a relatively short time (that may be considered as instantaneous) in each orbit the comet passes through the immediate vicinity of the sun and the planets; and at this instant the gravitational field of Jupiter (and to a lesser extent Saturn also) perturbs the cometary energy by a random component. Successive energy perturbations (in suitable units of energy) may be taken as independent random numbers η_1, η_2, ... drawn from a standardized normal distribution (2.3.20). A comet, starting with an energy $-z_0$, has accordingly energies

$$-z_0, \quad -z_1 = -z_0 + \eta_1, \quad -z_2 = -z_1 + \eta_2, \ldots \quad (4.2.1)$$

on successive orbits. This process continues until the first occasion on which z changes sign, whereupon the comet departs on a hyperbolic orbit and is lost from the solar system. According to Kepler's third law, the time taken to describe an orbit with energy $-z$ is $z^{-3/2}$ (in suitable units of time). Hence the total lifetime of the comet is

$$G = \sum_{i=0}^{T-1} z_i^{-3/2} \qquad (4.2.2)$$

where z_T is the first negative quantity in the sequence $z_0, z_1, \ldots$. Clearly, G is a random variable. The problem is to determine its distribution for a specified initial z_0. The index $-3/2$ in (4.2.2) renders this a very difficult theoretical problem. On the other hand, simulation is easy. In [8] the required η_i were generated by means of (3.4.17) with $n = 12$, and thence G was calculated from (4.2.1) and (4.2.2). This procedure was repeated N times, and the proportion $p(g)$ of those values of G (in this sample of size N) which did not exceed g gave a direct estimate of the required distribution function $P(G \leqslant g)$. In accordance with (2.4.14), the standard error of this estimate is

$$[p(g)\{1-p(g)\}/N]^{1/2}. \qquad (4.2.3)$$

In §1.1 we stated the general precept that the precision of Monte Carlo results may be improved by replacing a part of the random experiment (in this case the direct simulation) by mathematical theory. We can illustrate this precept in the present case. The probability that the comet will depart after only one orbit is

$$F = P(\eta_1 \geqslant z_0) = \int_{z_0}^{\infty} (2\pi)^{-1/2} e^{-t^2/2} dt, \qquad (4.2.4)$$

and this value of F is available from tables of the normal distribution. Accordingly we know that

$$1 - P(G \leqslant g) = \begin{cases} 1 & \text{if } g < z_0^{-3/2} \\ F & \text{if } g = z_0^{-3/2} \end{cases} \qquad (4.2.5)$$

and it remains to estimate $P(G > g) = 1 - P(G \leqslant g)$ for $g > z_0^{-3/2}$. When $g > z_0^{-3/2}$, we know that $T > 1$. Within the sample of size N, there is a subsample of size N^*, say, such that $T > 1$ for each calculation in the subsample. Let $1 - p^*(g)$ be the proportion of values of

G in this subsample exceeding g. Using the rules for conditional probability (§ 2.2), we see that $[1-p^*(g)]F$ is an estimator of $P(G > g)$, and that it has standard error

$$[p^*(g)\{1-p^*(g)\}/N^*]^{1/2} F, \qquad (4.2.6)$$

and this is smaller than (4.2.3) by a factor of

$$\left[1 - \frac{(1-F)\{1-p(g)\}}{Fp(g)}\right]^{1/2} \qquad (4.2.7)$$

approximately.

In direct simulation one sometimes needs to sample the basic random numbers from an unusual distribution in a way that matches the physical problem. We may have recourse to the kinds of device used in §3.4, or we may need to develop special techniques as in the following problem, arising in the field of metallurgy. If one has two concentric symmetrical bodies, cubes for instance, in random orientations, what is the distribution of the smallest angle through which one body must turn to come into coincidence with the other? This problem and others of the same nature were investigated by Mackenzie and Thomson [9] using a straightforward random sampling method. (The original problem has since been solved analytically [10], [11], but the Monte Carlo method remains the only effective one for some of the more complicated related problems.)

The interest of this calculation is centred on the generation of random orientations, represented by orthogonal 3×3 matrices. If the columns of such a matrix are the vectors $\mathbf{x}$, $\mathbf{y}$, and $\mathbf{z}$, then each of these vectors must have a uniform distribution over the surface of the unit sphere, but at the same time the three vectors must be mutually perpendicular. The method used by Mackenzie and Thomson is this. Let $x_1, x_2, x_3, y_1, y_2, y_3$ be independent standardized random normal deviates, and let

$$S = \sum x_r^2, \quad T = \sum y_r^2. \qquad (4.2.8)$$

Then

$$\mathbf{x} = [x_1, x_2, x_3]/S^{1/2}, \quad \mathbf{u} = [y_1, y_2, y_3]/T^{1/2} \qquad (4.2.9)$$

are independent uniformly-distributed unit vectors. Now take

$$\mathbf{y} = (\mathbf{u} - P\mathbf{x})/(1 - P^2)^{1/2}, \qquad (4.2.10)$$

where $P = \mathbf{x} \cdot \mathbf{u}$. Then $\mathbf{y}$ is a uniformly-distributed unit vector perpendicular to $\mathbf{x}$. Finally take $\mathbf{z} = \mathbf{x} \times \mathbf{y}$.

A computationally simpler procedure is to find four independent random normal deviates y_0, y_1, y_2, y_3 and to let $x_i = y_i / S^{1/2}$, where $S = \sum y_i^2$. Then we use

$$\begin{pmatrix} 1 - 2x_2^2 - 2x_3^2 & 2x_1 x_2 + 2x_0 x_3 & 2x_3 x_1 - 2x_0 x_2 \\ 2x_1 x_2 - 2x_0 x_3 & 1 - 2x_1^2 - 2x_3^2 & 2x_2 x_3 + 2x_0 x_1 \\ 2x_3 x_1 + 2x_0 x_2 & 2x_2 x_3 - 2x_0 x_1 & 1 - 2x_1^2 - 2x_2^2 \end{pmatrix} \quad (4.2.11)$$

as the required random orthogonal matrix. This procedure requires the calculation of only one square root, and it only needs 4 random normal deviates as opposed to the 6 of (4.2.8).

For examples of even more elaborate sampling techniques, devised for isotropic Gaussian processes, see [12] and [13].

4.3 Quantification of qualitative mathematical theory

Sometimes mathematical theory will provide the general form of a solution to a physical problem, while failing to supply the numerical values of certain constants occurring in the solution. The engineer's use of dimensional analysis is a familar case in point. For example, the drag T of a windmilling airscrew depends upon the air density ρ, the forward velocity V of the aircraft, and the diameter D of the circle swept by the airscrew. On the assumption that T is proportional to $\rho^a V^b D^c$, dimensional analysis gives $a = 1$, $b = c = 2$, so that

$$T = k\rho V^2 D^2, \quad (4.3.1)$$

where k is a numerical constant whose value can be determined from a physical experiment. (Even when engineers can determine the values of constants from theory, they are prone to check the theory against experiment and, very properly, to prefer the latter in any case of discrepancy.) In the same way, we may use Monte Carlo experiments to convert a qualitative mathematical formula into a quantitative one.

Consider, for example, the so-called travelling salesman problem. A travelling salesman, starting at and finally returning to his depot in a certain town, has to visit $n - 1$ other towns by way of the shortest possible route. If n is at all large it is prohibitively laborious to calculate the total mileage for each of the $(n-1)!$ possible orders of visiting the towns and to select the smallest. There are systematic computing algorithms for determining the optimum order and hence the value of l, the total length of the shortest route; but they require fairly large-scale computing resources. Let us see, therefore, whether we can find a simpler method. Clearly l will depend upon the total area A of the region containing the towns and upon the density n/A of the towns within this area. If we assume that l is proportional to $A^a(n/A)^b$, dimensional analysis shows that $a - b = \frac{1}{2}$. Further, if we multiply the area by a factor f while keeping the density constant, we shall multiply l by f. Hence $a = 1$; and we obtain

$$l = k(nA)^{1/2} \tag{4.3.2}$$

where k is a constant. In general, of course, l must depend upon the shape of the containing region and the detailed positions of the individual towns within it. However, it can be shown [14] that, except in a negligible fraction of particular cases (and this use of the word 'negligible' can be made mathematically precise) neither shape nor detailed positions matter provided n is reasonably large. Hence (4.3.2) provides us with a simple asymptotic formula for large n. To be able to use the formula we need to know the numerical value of k, and no known mathematical theory has yet been constructed to provide this. Instead, therefore, we quantify this general formula by means of a Monte Carlo experiment yielding the numerical value of k. We simply take a simple region (say a unit square), distribute n points in it uniformly at random and determine l and hence k. This may require large-scale computing to find l in the particular cases of the Monte Carlo experiment; but, once this has been done, we have calibrated the theory by means of the absolute constant k and can use (4.3.2) in further practical applications. It turns out that k is about 3/4.

4.4 Comparative simulation

Suppose that we wish to compare the results of two slightly different situations S_1 and S_2 in a given problem. We could, of course, simulate each situation independently and then compare the two sets of results. But it is much more satisfactory to use the *same* random numbers in the two situations, as we may from a supply of pseudorandom numbers generated in a prescribed fashion. For the difference between two unbiased estimates is an unbiased estimate of the difference, even when the estimates are dependent on one another, and the precision of the estimated difference will be greater if the dependence is such that, when the result in one situation happens by sampling variations to be overestimated, so is the result in the other situation by nearly the same amount. In brief, we do not wish a small difference between slightly different situations to be swamped by gross variations in each. This matter will be discussed further in §5.5 and §7.4.

Quite generally, good Monte Carlo practice uses each random number or each combination of random numbers several times over whenever it is safe to do so. It saves work and may increase precision. This applies especially in the more sophisticated types of Monte Carlo work in the subsequent chapters; see §8.3 and chapter 11 for examples.

CHAPTER 5

General Principles of the Monte Carlo Method

5.1 Introduction

Every Monte Carlo computation that leads to quantitative results may be regarded as estimating the value of a multiple integral. For suppose that no computation requires more than $N(= 10^{10}$ say) random numbers; then the results will be a (vector-valued) function

$$\mathbf{R}(\xi_1, \xi_2, \ldots, \xi_N) \qquad (5.1.1)$$

of the sequence of random numbers ξ_1, ξ_2, …. This is an unbiased estimator of

$$\int_0^1 \ldots \int_0^1 \mathbf{R}(x_1, \ldots, x_N)\, dx_1 \ldots dx_N. \qquad (5.1.2)$$

This way of looking at things is not always profitable, but the problem of evaluating integrals does provide a useful platform for exhibiting various Monte Carlo techniques of more general application. Initially, for the sake of simplicity, we shall take as our standard example the one-dimensional integral

$$\theta = \int_0^1 f(x)\, dx, \qquad (5.1.3)$$

despite the fact that such integrals can be evaluated far more efficiently by conventional numerical means than by Monte Carlo methods; and we suppose that $f \in L^2(0,1)$ (in other words, that $\int_0^1 [f(x)]^2\, dx$ exists, and therefore that θ exists). The extension to higher dimensions is sometimes obvious and sometimes rather difficult, depending on

50

the subtlety of the technique under discussion; this is in contrast to conventional numerical integration where the extension is nearly always difficult in computational practice.

We may define the relative efficiency of two Monte Carlo methods. Let the methods call for n_1 and n_2 units of computing time, respectively, and let the resulting estimates of θ have variances σ_1^2 and σ_2^2. Then the efficiency of method 2 relative to method 1 is

$$\frac{n_1 \sigma_1^2}{n_2 \sigma_2^2}. \qquad (5.1.4)$$

Notice that if we perform several independent computations by method 1 and average the results we do not change its efficiency. Often it suffices to take n_1 and n_2 in (5.1.4) to be the respective number of times that $f(.)$ is evaluated in each method. In any case, the efficiency ratio (5.1.4) is the product of two terms, the *variance ratio* σ_1^2/σ_2^2 and the *labour ratio* n_1/n_2. The former depends mainly on the problem and the Monte Carlo methods, and is easy to assess, at least in the examples discussed below; the latter depends partly on the Monte Carlo method and partly on the computing machinery available. Since we do not wish to burden this book with detailed considerations of computing resources, we shall only give rough and ready assessments of the labour ratio.

5.2 Crude Monte Carlo

If $\xi_1, \ldots, \xi_n$ are independent random numbers (rectangularly distributed between 0 and 1), then the quantities

$$f_i = f(\xi_i) \qquad (5.2.1)$$

are independent random variates with expectation θ. Therefore by (2.4.9)

$$f = \frac{1}{n} \sum_{i=1}^{n} f_i \qquad (5.2.2)$$

is an unbiased estimator of θ, and its variance is

$$\frac{1}{n} \int_0^1 (f(x) - \theta)^2 \, dx = \sigma^2/n. \qquad (5.2.3)$$

The standard error of f is thus

$$\sigma_{\bar{f}} = \sigma/\sqrt{n}. \qquad (5.2.4)$$

We shall refer to f as the *crude Monte Carlo* estimator of θ.

For example, take

$$f(x) = \frac{e^x - 1}{e - 1}, \qquad (5.2.5)$$

so that $\theta = 0.418$, $\sigma = 0.286$. We take 16 random numbers (actually extracted from a table of random numbers [1]) and evaluate (5.2.2). In this case we find that $f = 0.357$ so that $|f - \theta| = 0.061$, while the theoretical standard error is $\sigma/4 = 0.072$, in good agreement. The calculation is set out in Table 5.1.

In practice, we should probably not know the standard error, so that we should estimate it from the formula

$$s^2 = \frac{1}{n-1} \sum_{i=1}^{n} (f_i - f)^2, \qquad (5.2.6)$$

giving an estimate of $s = 0.29$ for σ, or 0.07 for $\sigma/\sqrt{n}$. We should then announce the result of our calculations as

$$\theta = 0.357 \pm 0.07, \qquad (5.2.7)$$

meaning that 0.357 is an observation from a distribution whose mean is θ and whose standard deviation we estimate at 0.07. Since, by the Central Limit Theorem we expect that the distribution of f is approximately normal, we may say with 95% confidence that we are within 2 standard deviations of the mean, i.e. that $0.22 \leqslant \theta \leqslant 0.50$. The phrase '$x$% confidence' signifies that with the foregoing rule of procedure repeatedly applied in actual or conceptual Monte Carlo experiments of a similar kind, x% of results would in the long run be correct. Unless we know the value of the estimand θ (in which case there would be no point in carrying out the Monte Carlo work, except for explanatory or demonstration purposes), we cannot say whether any particular result is correct or not. It is only because we know (from theory) that $\theta = 0.418$ that we also know the assertion '$0.22 \leqslant \theta \leqslant 0.50$' belongs to the 95% of correct assertions.

The factor of $\sqrt{n}$ in the denominator (5.2.4) implies that in order to halve the error we must take 4 times as many observations, and

Table 5.1

i	ξ_i	$f(\xi_i)$
1	0·96	0·938
2	0·28	0·188
3	0·21	0·136
4	0·94	0·908
5	0·35	0·244
6	0·40	0·286
7	0·10	0·061
8	0·52	0·397
9	0·18	0·115
10	0·08	0·048
11	0·50	0·378
12	0·83	0·753
13	0·73	0·626
14	0·25	0·165
15	0·33	0·228
16	0·34	0·236
Average		0·357

so on, so that, in our example, in order to achieve 2 significant figures of accuracy (with a standard error less than 0·005) we should need to make about 3000 observations of values of f.

We shall see later how we may drastically cut down this number, but, first of all, let us compare this method with one even less efficient, namely *hit-or-miss Monte Carlo*.

Suppose that $0 \leqslant f(x) \leqslant 1$ when $0 \leqslant x \leqslant 1$. Then we may draw the curve $y = f(x)$ in the unit square $0 \leqslant x, y \leqslant 1$, and θ is the proportion of the area of the square beneath the curve. To put the matter formally, we may write

$$\left. \begin{array}{l} f(x) = \int\limits_0^1 g(x,y)\,dy, \\ g(x,y) = 0 \text{ if } f(x) < y \\ = 1 \text{ if } f(x) \geqslant y. \end{array} \right\} \quad (5.2.8)$$

where

We may then estimate θ as a double integral,

$$\theta = \int\limits_0^1 \int\limits_0^1 g(x, y)\,dx\,dy, \qquad (5.2.9)$$

by the estimator

$$\bar{g} = \frac{1}{n} \sum_{i=1}^{n} g(\xi_{2i-1}, \xi_{2i}) = \frac{n^*}{n}, \qquad (5.2.10)$$

where n^* is the number of occasions on which $f(\xi_{2i-1}) \geqslant \xi_{2i}$. In other words, we take n points at random in the unit square, and count the proportion of them which lie below the curve $y = f(x)$. This is sampling from the binomial distribution (2.3.19) with $p = \theta$, and the standard error is

$$\sqrt{\frac{\theta(1-\theta)}{n}}. \qquad (5.2.11)$$

In our example, this is $0 \cdot 123$. Compared with crude Monte Carlo, the hit-or-miss method has a variance ratio of $(0 \cdot 072/0 \cdot 123)^2 = 0 \cdot 34$. If we take the labour ratio as 1, the efficiency ratio also is $0 \cdot 34$. In other words, to achieve a result of given accuracy, hit-or-miss Monte Carlo calls for nearly 3 times as much sampling as crude Monte Carlo. The factor 3 depends upon the problem; but hit-or-miss is always worse than even crude Monte Carlo [see (5.2.14)]. It is worth stressing this point because, historically, hit-or-miss methods were once the ones most usually propounded in explanation of Monte Carlo techniques; they were, of course, the easiest methods to understand, particularly if explained in the kind of graphical language involving a curve in a square, and in those days there was little thought given to efficiency. Herein lay one of the causes of the bad name acquired by Monte Carlo methods in the 1950's (see § 1.2).

The comparison between hit-or-miss and crude Monte Carlo methods also illustrates a general principle of Monte Carlo work: *if, at any point of a Monte Carlo calculation, we can replace an estimate by an exact value, we shall reduce the sampling error in the final result.* In the present instance, the variance in hit-or-miss (i.e. binomial) sampling is

$$\sigma_B^2 = \theta(1-\theta)/n, \qquad (5.2.12)$$

and in crude Monte Carlo

$$\sigma_C^2 = \frac{1}{n} \int_0^1 (f-\theta)^2 \, dx = \frac{1}{n} \int_0^1 f^2 \, dx - \theta^2/n. \qquad (5.2.13)$$

Thus

$$\sigma_B^2 - \sigma_C^2 = \frac{\theta}{n} - \frac{1}{n} \int_0^1 f^2 \, dx = \frac{1}{n} \int_0^1 f(1-f) \, dx > 0, \qquad (5.2.14)$$

reflecting the fact that the passage from hit-or-miss to crude sampling is equivalent to replacing $g(x, \xi)$ by its expectation $f(x)$.

5.3 Stratified sampling

In stratified sampling, we break the range of integration into several pieces, say $\alpha_{j-1} < x \leqslant \alpha_j$ where $0 = \alpha_0 < \alpha_1 < \ldots < \alpha_k = 1$, and apply crude Monte Carlo sampling to each piece separately. The estimator of θ is then of the form

$$t = \sum_{j=1}^k \sum_{i=1}^{n_j} (\alpha_j - \alpha_{j-1}) \frac{1}{n_j} f(\alpha_{j-1} + (\alpha_j - \alpha_{j-1}) \xi_{ij}) \qquad (5.3.1)$$

when we decide (beforehand) to sample n_j points from the jth piece. The estimator (5.3.1) is unbiased and its variance is

$$\sigma_t^2 = \sum_{j=1}^k \frac{(\alpha_j - \alpha_{j-1})}{n_j} \int_{\alpha_{j-1}}^{\alpha_j} f(x)^2 \, dx - \sum_{j=1}^k \frac{1}{n_j} \left\{ \int_{\alpha_{j-1}}^{\alpha_j} f(x) \, dx \right\}^2. \qquad (5.3.2)$$

This variance may be less than σ_f^2, with $n = \sum n_j$, if the stratification is well carried out so that the differences between the mean values of f in the various pieces are greater than the variations of f within the pieces. When the stratification points are prescribed, the best way of distributing the sample points among the strata is so that n_j^2 is proportional to

$$\left[(\alpha_j - \alpha_{j-1}) \int_{\alpha_{j-1}}^{\alpha_j} f(x)^2 \, dx - \left\{ \int_{\alpha_{j-1}}^{\alpha_j} f(x) \, dx \right\}^2 \right]. \qquad (5.3.3)$$

There are various ways of choosing the α_j [2]. The simplest is to divide the original interval into k equal pieces, $\alpha_j = j/k$. A better way is to choose the α_j so that the variation of f is the same in each piece.

For example, in Table 5.2 we integrate the function (5.2.5) by taking 4 points from each of 4 strata, the divisions being chosen in each of the above ways. The resulting estimates of θ are 0·399 and 0·409 respectively, so that $|f - \theta| = 0.019$ and 0·009, both of these being improvements on the crude estimate. Theoretically the stratified estimators

Table 5.2

	(1) $\alpha_j =$ 0·25 0·50 0·75		(2) $\alpha_j =$ 0·36 0·62 0·83	
i	ξ_i	$f(\xi_i)$	ξ_i	$f(\xi_i)$
1	0·24	0·158	0·35	0·244
2	0·07	0·042	0·10	0·061
3	0·05	0·030	0·08	0·048
4	0·24	0·158	0·34	0·236
5	0·34	0·236	0·45	0·331
6	0·35	0·244	0·46	0·340
7	0·27	0·180	0·39	0·277
8	0·38	0·269	0·49	0·368
9	0·54	0·417	0·66	0·544
10	0·52	0·397	0·64	0·521
11	0·63	0·511	0·73	0·626
12	0·71	0·602	0·79	0·700
13	0·93	0·893	0·95	0·923
14	0·81	0·726	0·87	0·807
15	0·83	0·752	0·89	0·835
16	0·84	0·766	0·89	0·835
Estimate of θ:		0·399		0·409

have standard errors of 0·0197 and 0·0193 respectively. Compared with crude Monte Carlo the variance ratios are $(0·072/0·0197)^2 = 13$ and $(0·072/0·0193)^2 = 14$ respectively. If we suppose that stratification involves 30% or 40% more labour, the labour ratios are $1/1·3$ and $1/1·4$. We conclude that stratified sampling is here about 10 times as efficient as crude Monte Carlo. As a general rough working rule, the efficiency of stratified sampling increases as the square of the number of strata.

To estimate the standard error of these results, we must replace (5.3.2) by an estimate depending on the sample itself, namely

$$s_t^2 = \sum_{j=1}^{k} \frac{(\alpha_j - \alpha_{j-1})^2}{n_j(n_j - 1)} \sum_{i=1}^{n_j} (f_{ij} - f_j)^2, \qquad (5.3.4)$$

where

$$\left.\begin{array}{l} f_{ij} = f(\alpha_{j-1} + (\alpha_j - \alpha_{j-1})\, \xi_{ij}), \\[2mm] f_j = \dfrac{1}{n_j} \displaystyle\sum_{i=1}^{n_j} f_{ij}. \end{array}\right\} \qquad (5.3.5)$$

In the present instance, we find that $s_t^2 = 0·00032$ and $0·00049$ in the respective cases. This gives us s_t, and we may quote the estimates and their standard errors in the form

$$0·399 \pm 0·018 \quad \text{and} \quad 0·409 \pm 0·022, \qquad (5.3.6)$$

or in the form of confidence intervals in the manner following (5.2.7).

5.4 Importance sampling

We have

$$\theta = \int_0^1 f(x)\,dx = \int_0^1 \frac{f(x)}{g(x)} g(x)\,dx = \int_0^1 \frac{f(x)}{g(x)} dG(x), \qquad (5.4.1)$$

for any functions g and G satisfying

$$G(x) = \int_0^x g(y)\,dy. \qquad (5.4.2)$$

Let us restrict g to be a positive-valued function such that

$$G(1) = \int_0^1 g(y)\,dy = 1. \tag{5.4.3}$$

Then $G(x)$ is a distribution function for $0 \le x \le 1$, and, if η is a random number sampled from the distribution G, (5.4.1) shows that $f(\eta)/g(\eta)$ has expectation θ and variance

$$\sigma_{f/g}^2 = \int_0^1 \left(\frac{f(x)}{g(x)} - \theta\right)^2 dG(x). \tag{5.4.4}$$

The object in importance sampling is to concentrate the distribution of the sample points in the parts of the interval that are of most 'importance' instead of spreading them out evenly. So as not to bias the result, we compensate for thus distorting the distribution by taking f/g in place of f as our estimator.

We notice that if f is also positive-valued we can take g to be proportional to f; $g = cf$, say. Then (5.4.1) and (5.4.3) imply $c = 1/\theta$; whereupon (5.4.4) yields $\sigma_{f/g}^2 = 0$. We thus appear to have a perfect Monte Carlo method, giving the exact answer every time. This method is unfortunately useless, since to sample f/g we must know g, and to determine $g(=f/\theta)$ we must know θ, and if we already know θ we do not need Monte Carlo methods to estimate it.

Not all is lost, however. We notice first that we always get an unbiased estimate of θ, whatever positive function g we use. Our object is to select some g to reduce the standard error of our estimate. This estimate is an average of observed values of f/g, and it will have a small sampling variance if f/g is as constant as we can make it; we cannot make it wholly constant for the reasons just stated. We therefore want g to mimic f on the one hand, so that the ratio f/g will vary little; but on the other hand we have to restrict our choice of g to functions that we can integrate theoretically, because we must satisfy (5.4.3). These are conflicting requirements: g must be simple enough for us to know its integral theoretically, whereas f is so complicated that it eludes theoretical integration and requires Monte Carlo estimation. We must therefore compromise between these requirements: and a good compromise should yield an estimator of θ with a substantially smaller standard error than the crude Monte Carlo estimator.

Let us imagine that our standard integrand (5.2.5) is beyond our powers of theoretical integration, but that we do know how to integrate the function x between 0 and 1. Both functions increase over $0 \leqslant x \leqslant 1$, and therefore mimic one another to some extent. Indeed their relative mimicry is about as much as one may normally hope for in practice. If we take $g(x) = x$, we find from (5.4.4) $\sigma_{f/g}^2 = 0 \cdot 00274$. The corresponding variance (for a single observation) in crude Monte Carlo is $0 \cdot 0820$; so that the variance ratio is $29 \cdot 9$. Assuming a labour ratio of $1/3$ (i.e. three times as much work in the importance sampling), we have an efficiency gain of 10.

The transformation (5.4.1) is especially relevant for unbounded integrands. It is clear that one should prefer the distribution whose density function follows most closely the shape of f and makes f/g bounded.

5.5 Control variates

The sampling error in the crude Monte Carlo estimate of θ in (5.2.2) arises from the variation of $f(x)$ as x runs over $0 \leqslant x \leqslant 1$. Importance sampling (§ 5.4) and control variates [3] are two different techniques for reducing this variation and hence improving efficiency of estimation. For control variates, we break (5.1.3) into two parts,

$$\theta = \int\limits_0^1 \phi(x)\,dx + \int\limits_0^1 [f(x) - \phi(x)]\,dx, \qquad (5.5.1)$$

which we integrate separately, the first by mathematical theory and the second by crude Monte Carlo. Thus ϕ must be a simple enough function to integrate theoretically. On the other hand, unless ϕ mimics f and absorbs most of its variation, the Monte Carlo errors in the second integral will not be appreciably smaller than those in the original crude Monte Carlo. Once again we have two conflicting requirements to compromise between. We call $\phi(\xi)$ the *control variate* for $f(\xi)$. The whole method is simply another example of the precept in § 1.1 that one should as far as possible replace Monte Carlo experiment by theoretical analysis.

If we choose $\phi(x) = x$ in connexion with our standard example (5.2.5), we find that $\sigma_{f-\phi}^2 = 0.001358$, as against the original $\sigma_f^2 = 0.08198$. The control variate method gives a variance ratio of 60·4 compared with the crude method in this case. It requires about twice as much labour, so the efficiency is about 30 times that of crude Monte Carlo.

There are various other ways of looking at this method. For instance, when estimating an unknown parameter θ by means of an estimator t, we may seek another estimator t', which has a strong positive correlation with t, and whose expectation is a numerically known quantity θ'. We then sample t and t' simultaneously, and use $t - t' + \theta'$ as the estimator of θ. In the present case

$$t = \frac{1}{n} \sum_{i=1}^{n} f(\xi_i), \quad t' = \frac{1}{n} \sum_{i=1}^{n} \phi(\xi_i), \qquad (5.5.2)$$

and θ' is the first integral in (5.5.1). We use the *same* random numbers ξ_i in both equations of (5.5.2) to produce the required positive correlation. We have

$$\mathrm{var}\,(t - t' + \theta) = \mathrm{var}\,t + \mathrm{var}\,t' - 2\,\mathrm{cov}\,(t, t'), \qquad (5.5.3)$$

and this will be less than $\mathrm{var}\,t$ if $2\,\mathrm{cov}(t, t') > \mathrm{var}\,t'$.

Sometimes we have to find a Monte Carlo solution of a complicated problem, of which a simpler version is amenable to theoretical analysis. In this case we attack both versions simultaneously by Monte Carlo Methods, using identical random numbers; in the foregoing notation, t and t' are the Monte Carlo estimates for the complicated and the simpler versions respectively, and θ' is the known analytical solution of the simpler problem. In Chapter 7, we shall meet an example of this; the complicated problem is the penetrability of a finite neutron shield, while the simpler version with a known analytical solution is concerned with the corresponding infinite shield.

5.6 Antithetic variates

In the control variate method, we sought a second estimate t' having a known expectation and a strong positive correlation with the original estimator t. In the antithetic variate method, we seek an estimator t'', having the same (unknown) expectation as t and a strong negative correlation with t. Then $\frac{1}{2}(t+t'')$ will be an unbiased estimator of θ, and its sampling variance

$$\text{var}\left[\tfrac{1}{2}(t+t'')\right] = \tfrac{1}{4}\text{var}\,t + \tfrac{1}{4}\text{var}\,t'' + \tfrac{1}{2}\text{cov}\,(t,t''), \qquad (5.6.1)$$

in which $\text{cov}(t,t'')$ is negative, can sometimes be made smaller than $\text{var}\,t$ by suitably selecting t''.

For example, $1-\xi$ is rectangularly distributed whenever ξ is, so that $f(\xi)$ and $f(1-\xi)$ are both unbiased estimators of θ. When f is a monotone function, $f(\xi)$ and $f(1-\xi)$ will be negatively correlated. Thus we could take

$$\tfrac{1}{2}(t+t'') = \tfrac{1}{2}f(\xi) + \tfrac{1}{2}f(1-\xi) \qquad (5.6.2)$$

as an estimator of θ. For our standard example (5.2.5), the variance of (5.6.2) is 0·00132, giving a variance ratio of 62 in comparison with crude Monte Carlo. The labour ratio will be about $\frac{1}{2}$, since we have twice as many evaluations of f (but only the same number of random numbers to generate). Thus the efficiency gain is about 31.

From the practical viewpoint, the mathematical conditions that a Monte Carlo technique has to satisfy govern its efficiency. As in the case of importance sampling and control variates, we are usually unable to satisfy the conditions in the theoretically optimum way, and we have to be content with some compromise. When the conditions are fairly loose and flexible, it is easier to reach a good compromise. This is the case with antithetic variates; in practice it is relative easy to find negatively correlated unbiased estimators of θ, usually easier than it is to find an equally satisfactory control variate or importance function. Accordingly, the antithetic variate method tends to be more efficient in practice. We use the term *antithetic variates* to describe any set of estimators which mutually compensate each other's variations.

The name, if not the idea (see [4]), of antithetic variates was introduced by Hammersley and Morton in 1956 [5]. It is based on the following theorem (stated in [6], and proved there and in [7]):

Theorem. If I denotes the infimum of $\text{var}\sum\limits_{j=1}^{n} g_j(\xi_j)$ *when all possible stochastic or functional dependences between the ξ_j are considered,* *subject to each ξ_j being rectangularly distributed between $(0,1)$, then, provided that g_j are bounded† functions,*

$$\inf_{x_j \,\epsilon\, \mathfrak{X} \,(j\,=\,1,\,2,\,\ldots,\,n)} \mathrm{var}\left\{\sum_{j=1}^{n} g_j[x_j(\xi)]\right\} = I, \quad (5.6.3)$$

where $\mathfrak{X}$ denotes the class of functions $x(z)$ with the properties (i) $x(z)$ *is a* (1,1) *mapping of the interval* $(0,1)$ *onto itself, and* (ii) *except at at most a finite number of points z, $dx/dz = 1$.*

In other words, whenever we have an estimator consisting of a sum of random variables, it is possible to arrange for there to be a strict functional dependence between them, such that the estimator remains unbiased, while its variance comes arbitrarily close to the smallest that can be attained with these variables. Essentially, we 'rearrange' the random variables (in the sense of Chapter X of [8]) by permuting finite subintervals, in order to make the sum of the rearranged functions as nearly constant as possible. In the case of two monotone functions, this is done by rearranging them so that one is monotone increasing and the other decreasing; this is what happened when we applied (5.6.2) to the function (5.2.5).

The systems of antithetic variates treated in [5] are based on a stratification of the interval. If we take $k = 2$, $\alpha_1 = \alpha$ and $n_j = n$ in (5.3.1), we have

$$t = \frac{1}{n} \sum_{i=1}^{n} \{\alpha f(\alpha \xi_{i1}) + (1-\alpha) f[\alpha + (1-\alpha)\,\xi_{i2}]\}. \quad (5.6.4)$$

We may now introduce the simple dependence $\xi_{i2} = \xi_{i1} = \xi_i$, leading to

$$t = \frac{1}{n} \sum_{i=1}^{n} \{\alpha f(\alpha \xi_i) + (1-\alpha) f[\alpha + (1-\alpha)\,\xi_i]\} = \frac{1}{n} \sum_{i=1}^{n} \mathfrak{S}_\alpha f(\xi_i),$$
$$(5.6.5)$$

or, alternatively, $1 - \xi_{i2} = \xi_{i1} = \xi_i$, leading to

$$t = \frac{1}{n} \sum_{i=1}^{n} \{\alpha f(\alpha \xi_i) + (1-\alpha) f[1 - (1-\alpha)\,\xi_i]\} = \frac{1}{n} \sum_{i=1}^{n} \mathfrak{T}_\alpha f(\xi_i).$$
$$(5.6.6)$$

† The condition that the g_j are bounded may be unnecessary, but this has never been established.

The transformations $\mathfrak{S}_\alpha$ and $\mathfrak{T}_\alpha$ are linear, preserve expectations, and double the number of times that the function has to be evaluated. One can see that the transformation (5.6.2) is essentially equivalent to $\mathfrak{T}_{1/2}$.

If the function f is monotone, we are naturally tempted to try the effect of $\mathfrak{T}_\alpha$, since one of its two terms is increasing and the other is decreasing. We find

$$\text{var}\,(\mathfrak{T}_\alpha f(\xi)) = \int_0^1 \{\alpha f(\alpha x) + (1-\alpha)f[1-(1-\alpha)x]\}^2\,dx - \theta^2$$

$$= \alpha \int_0^\alpha f(x)^2\,dx + (1-\alpha)\int_\alpha^1 f(x)^2\,dx - \theta^2 +$$

$$+ 2(1-\alpha)\int_0^\alpha f(x)f[1-(\alpha^{-1}-1)x]\,dx, \quad (5.6.7)$$

which has a minimum at some value of α between 0 and 1. It is difficult to locate this value exactly, on account of the complexity of the formula, but an adequate rule of thumb is to choose α to make $\mathfrak{T}_\alpha f(0) = \mathfrak{T}_\alpha f(1)$, or to find the root of

$$f(\alpha) = (1-\alpha)f(1) + \alpha f(0). \quad (5.6.8)$$

In the case of (5.2.5), the solution of (5.6.8) is $\alpha = 0\cdot5615$, giving a standard deviation of $(0\cdot000083)^{1/2} = 0\cdot009$, representing a gain of efficiency of 490 over crude Monte Carlo. In this case the rule of thumb gives a very good approximation to the minimum.

We must emphasize, however, that an error in determining α is never catastrophic, since the resulting estimator of θ will remain completely unbiased, and the only ill effect may be that we lose some efficiency.

It is rather more difficult to give rules for applying $\mathfrak{S}_\alpha$ and $\mathfrak{T}_\alpha$ to general functions.

Another useful transformation is given by

$$\mathfrak{U}_m f(\xi) = \frac{1}{m}\sum_{j=0}^{m-1} f\left(\frac{\xi+j}{m}\right). \quad (5.6.9)$$

If f is a periodic† function with period 1, then

$$\text{var}\{\mathfrak{U}_m f(\xi)\} = O(e^{-km}) \qquad \text{as } m \to \infty, \qquad (5.6.10)$$

where the extension of f to the complex plane is regular in the strip $-k < 4\pi\Im(z) < k$. Otherwise there is an asymptotic expansion, based on the Euler-Maclaurin summation formula,

$$\text{var}\{\mathfrak{U}_m f(\xi)\} = \sum_{r,s \geqslant 0} \frac{(-)^r \Delta_r \Delta_s B_{r+s+2}}{(r+s+2)! \, m^{r+s+2}} = \frac{\Delta_0^2}{12m^2} + \frac{\Delta_1^2 - 2\Delta_0\Delta_2}{720m^4} +$$

$$+ \frac{\Delta_2^2 - 2\Delta_1\Delta_3 + 2\Delta_0\Delta_4}{30240m^6} + o(m^{-6}),$$

$$(5.6.11)$$

where B_m are the Bernoulli numbers and $\Delta_j = f^{(j)}(1) - f^{(j)}(0)$. It is necessary that f should be continuous and that all the derivatives used should also be continuous throughout the interval $(0, 1)$.

This indicates that if f is made to be such that $\Delta_0 = \Delta_1 = \ldots = \Delta_M = 0$ while $f^{(M+1)}$ and $f^{(M+2)}$ exist and are continuous, then

$$\text{var}\{\mathfrak{U}_m f(\xi)\} = o(m^{-2(M+1)}). \qquad (5.6.12)$$

Comparing this with crude Monte Carlo, where

$$\text{var}\left\{\frac{1}{m}\sum_{i=1}^m f(\xi)\right\} = O(m^{-1}), \qquad (5.6.13)$$

we see that, under the stated conditions on f, the antithetic-variate method is enormously more efficient, provided that m is large enough.

The condition $\Delta_0 = 0$ is easily achieved by carrying out the transformation $\mathfrak{T}_\alpha$, with α given by (5.6.8), or more simply by the transformation (5.6.2). Further Δ's may be reduced to zero, without requiring us to know anything about f except that enough derivatives exist,

† For another method of integrating periodic functions, see the paper by Haselgrove [9]. A graphic illustration [5] of the application of $\mathfrak{U}_m$ to a periodic function is given by the improvement of Buffon's needle experiment, in which the single needle is replaced by a regular polygon of $2m$ sides.

by various transformations discussed in [10], of which the first, eliminating Δ_1, is

$$\frac{2}{3}F\left(\frac{\xi}{2}\right)+\frac{2}{3}F\left(\frac{1+\xi}{2}\right)-\frac{1}{3}F(\xi), \qquad (5.6.14)$$

where $$F(\xi) = \mathfrak{T}_\alpha f(\xi) \qquad (5.6.15)$$

and α is given by (5.6.8).

It is instructive to summarize the performances of the several variance-reducing techniques discussed in this chapter. Table 5.3 gives, for the standard example (5.2.5), the variance ratio, the labour ratio, and the resulting efficiency of the different techniques, each in comparison with crude Monte Carlo. The figures for the labour ratio

Table 5.3

Method and defining equation	Variance ratio	Labour ratio	Efficiency gain
Hit-or-miss (5.2.10)	0·34	1/1	0·34
Stratified sampling, 4 equal strata (5.3.1)	13	1/1·3	10
Importance sampling (5.4.1), $g(x) = x$	29·9	1/3	10
Control variate (5.5.1), $\phi(x) = x$	60·4	1/2	30
Antithetic variate (5.6.2)	62	1/2	31
Antithetic $\mathfrak{T}_\alpha$ (5.6.6) and (5.6.8)	985	1/2	490
Antithetic $\mathfrak{U}_2\mathfrak{T}_\alpha$ (5.6.9) and (5.6.8)	$1\cdot56 \times 10^4$	1/4	3900
Antithetic $\mathfrak{U}_4\mathfrak{T}_\alpha$ (5.6.9) and (5.6.8)	$2\cdot49 \times 10^5$	1/8	31000
Antithetic $\mathfrak{U}_8\mathfrak{T}_\alpha$ (5.6.9) and (5.6.8)	$3\cdot98 \times 10^6$	1/16	250000
Antithetic (5.6.14), (5.6.15) and (5.6.8)	$2\cdot95 \times 10^6$	1/6	460000
Orthonormal (5.8.15)	$7\cdot2 \times 10^5$	1/3	240000

are rather rough and ready, for the reasons already explained, but they are of the right order of magnitude, and so too are the resulting figures for efficiency. Some of the later entries in this table are taken from [5] and [10], where some even more powerful antithetic transformations appear as well.

Various attempts [5], [10], [11] have been made to extend the antithetic variate method to multiple integrals, but the resulting formulae are clumsy, difficult to manage, and of doubtful general utility. The method will work for multiple integrals if a good deal of detailed care and attention is paid to the behaviour of the integrand, with *ad hoc* antithetic transformations to meet this behaviour. This can be done for paper-and-pencil Monte Carlo work, but is unsatisfactory for calculations on an electronic computer. An example of pencil-and-paper application of antithetic variates to a six-dimensional integral, leading to an efficiency gain of about 160, occurs in [5]. The simplest forms of antithetic transformations have proved rather more successful in nuclear reactor calculations; an efficiency gain of about 10 can often be secured by making a pair of neutrons emerge from a collision in diametrically opposite directions, and this device can be satisfactorily programmed on a high-speed computer provided it is used sparingly and at judicious points in the computation. If used unsparingly, it produces an unmanageably large number of neutron tracks.

5.7 Regression methods

Suppose that we have several unknown estimands $\theta_1, \theta_2, \ldots, \theta_p$ and a set of estimators $t_1, t_2, \ldots, t_n (n \geqslant p)$ with the property that

$$\mathscr{E}t_i = x_{i1}\theta_1 + x_{i2}\theta_2 + \ldots + x_{ip}\theta_p \quad (i = 1, 2, \ldots, n), \quad (5.7.1)$$

where the x_{ij} are a set of *known* constants. The antithetic variate technique of §5.6 is a particular case of this situation; for instance, if $t_1 = t$ and $t_2 = t''$ as in (5.6.1), and $\theta_1 = \theta$, we have $\mathscr{E}t_1 = \mathscr{E}t_2 = \theta_1$; this is the situation described by (5.7.1) with $p = 1, n = 2, x_{11} = x_{21} = 1$, and (of course) 1 is a known constant.

The equation (5.7.1) is an example of the situation discussed in § 2.6 if we identify the estimands $\theta_1, \ldots, \theta_p$ with the regression coefficients $\beta_1, \ldots, \beta_p$. According to (2.6.1), the minimum-variance unbiased linear estimator of $\theta = \{\theta_1, \ldots, \theta_p\}$ will be

$$t^* = (X'V^{-1}X)^{-1}X'V^{-1}t, \quad (5.7.2)$$

where X is the $n \times p$ matrix (x_{ij}), where V is the $n \times n$ variance-covariance matrix of the t_i's, and where $t = \{t_1, \ldots, t_n\}$. Here everything is known except perhaps $\mathbf{V}$. Suppose that we consider the alternative estimator

$$\mathbf{t}_0^* = (\mathbf{X}'\mathbf{V}_0^{-1}\mathbf{X})^{-1}\mathbf{X}'\mathbf{V}_0^{-1}\mathbf{t}, \qquad (5.7.3)$$

where $\mathbf{V}_0$ is some other variance-covariance matrix. Because (5.7.1) takes the form

$$\mathscr{E}\mathbf{t} = \mathbf{X}\boldsymbol{\theta} \qquad (5.7.4)$$

when written in matrix notation, and because $\mathbf{t}_0^*$ is a linear function of $\mathbf{t}$, we have

$$\mathscr{E}\mathbf{t}_0^* = \mathscr{E}(\mathbf{X}'\mathbf{V}_0^{-1}\mathbf{X})^{-1}\mathbf{X}'\mathbf{V}_0^{-1}\mathbf{t} = (\mathbf{X}'\mathbf{V}_0^{-1}\mathbf{X})^{-1}\mathbf{X}'\mathbf{V}_0^{-1}\mathscr{E}\mathbf{t}$$
$$= (\mathbf{X}'\mathbf{V}_0^{-1}\mathbf{X})^{-1}\mathbf{X}'\mathbf{V}_0^{-1}\mathbf{X}\boldsymbol{\theta} = \boldsymbol{\theta}. \qquad (5.7.5)$$

Hence, whatever $\mathbf{V}_0$ we use in (5.7.3), $\mathbf{t}_0^*$ is an unbiased estimator of $\boldsymbol{\theta}$. It will not be a minimum-variance estimator if $\mathbf{V}_0 \neq \mathbf{V}$; but if $\mathbf{V}_0$ is a reasonable approximation to $\mathbf{V}$, then $\mathbf{t}_0^*$ will have a very nearly minimum variance, particularly since first-order deviations in x in the neighbourhood of a minimum of a function $F(x)$ only cause second-order variations in $F(x)$. Thus if $\mathbf{V}$ is unknown, we may replace it by an estimate $\mathbf{V}_0$. As stated in § 2.6, the sampling variance-covariance matrix of $\mathbf{t}^*$ is

$$\operatorname{var}\mathbf{t}^* = (\mathbf{X}'\mathbf{V}^{-1}\mathbf{X})^{-1}, \qquad (5.7.6)$$

and to the second order of small quantities this will also be the sampling variance-covariance matrix of $\mathbf{t}_0^*$. The formula

$$\operatorname{var}\mathbf{t}_0^* = (\mathbf{X}'\mathbf{V}_0^{-1}\mathbf{X})^{-1} \qquad (5.7.7)$$

will therefore contain first-order errors, but these constitute first-order errors in the first-order standard error of $\mathbf{t}_0^*$, and therefore lead only to second-order errors in assigning a confidence interval to $\boldsymbol{\theta}$.

What we do in practice, therefore, is to calculate N independent sets of estimates $t_1, t_2, \ldots, t_n$ which we may denote by $t_{1k}, t_{2k}, \ldots, t_{nk}$ ($k = 1, 2, \ldots, N$), from which we can estimate v_{ij} by means of

$$v_{ij^0} = \sum_{k=1}^{N} (t_{ik}-\bar{t}_i)(t_{jk}-\bar{t}_j)/(N-1), \qquad (5.7.8)$$

where

$$\bar{t}_i = \sum_{k=1}^{N} t_{ik}/N. \qquad (5.7.9)$$

We define $\mathbf{V}_0$ to be the matrix (v_{ij0}), and employ the estimator

$$\mathbf{t}_0^* = (\mathbf{X}' \mathbf{V}_0^{-1} \mathbf{X})^{-1} \mathbf{X}' \mathbf{V}_0^{-1} \bar{\mathbf{t}} \qquad (5.7.10)$$

as an estimator of $\mathbf{\theta}$, where $\bar{\mathbf{t}} = \{\bar{t}_1, \ldots, \bar{t}_n\}$. It is a nearly unbiased estimator whose sampling variance-covariance matrix is

$$(\mathbf{X}' \mathbf{V}_0^{-1} \mathbf{X})^{-1}/N, \qquad (5.7.11)$$

apart from errors of order $N^{-3/2}$.

When we are dealing with n antithetic estimators of a single parameter $\theta = \theta_1$, the matrix $\mathbf{X}$ reduces to a single column $\mathbf{x}$. In many particular cases each element of $\mathbf{x}$ will be 1.

Let us consider an example. The estimators

$$\left.\begin{array}{l} t_1 = \tfrac{1}{2}f(\xi) + \tfrac{1}{2}f(1-\xi) \\ t_2 = \tfrac{1}{4}f(\tfrac{1}{2}\xi) + \tfrac{1}{4}f(\tfrac{1}{2} - \tfrac{1}{2}\xi) + \tfrac{1}{4}f(\tfrac{1}{2} + \tfrac{1}{2}\xi) + \tfrac{1}{4}f(1 - \tfrac{1}{2}\xi) \end{array}\right\} \quad (5.7.12)$$

are both unbiased estimators of the integral θ defined by (5.1.3). Thus we have the particular case $p = 1$, $n = 2$, $\mathbf{x} = \{1,1\}$. We took $N = 100$ for the standard integral (5.2.5), whose correct numerical value is 0·4180227. From our sample of 100 we found the estimates $\bar{t}_1 = 0·4218353$, $\bar{t}_2 = 0·4189959$, which are both overestimates. The sample also gave

$$v_{110} = 0·00131493, \quad v_{220} = 0·0008509, \quad v_{120} = v_{210} = 0·00033449.$$

From (5.7.3) we then computed $t_0^* = 0·4180273$, which is much closer to the true answer than either $\bar{t}_1$ or $\bar{t}_2$. The value of (5.7.11) was $0·000\,000\,000\,094 = (0·000\,009\,7)^2$. We can summarize the estimates and their standard errors for the three methods as follows:

$$\left.\begin{array}{l} \bar{t}_1 = 0·4218 \pm 0·0036 \\ \bar{t}_2 = 0·4190 \pm 0·0029 \\ t_0^* = 0·4180273 \pm 0·000\,009\,7 \\ \theta = 0·4180227 \text{ (exact)}. \end{array}\right\} \qquad (5.7.13)$$

As we might well hope, no estimate differs from the exact result by more than twice its standard error. The gain in precision from the comparatively slight extra labour in calculating t_0^* is remarkable.

It is interesting to compare the regression method with a pure antithetic variate method. One of the estimators arising in [10] takes the form

$$\tfrac{4}{3}t_2 - \tfrac{1}{3}t_1, \qquad (5.7.14)$$

whereas the regression method has in effect led us (for this particular function) to the estimator

$$1 \cdot 3411 t_2 - 0 \cdot 3411 t_1. \qquad (5.7.15)$$

These two estimators are remarkably similar. However (5.7.14) in fact leads to the inferior estimate $0 \cdot 41804 \pm 0 \cdot 000022$.

The attractions of regression methods, which can of course be applied in a wide variety of different Monte Carlo situations, are these:

  (i) they introduce little, if any, bias to the answer;
 (ii) if applied in a situation where correlation exists, they exploit this correlation and thereby reduce the errors in the final estimates, sometimes by a very considerable factor;
(iii) if unwittingly applied in a situation where no correlation exists, they do not lessen the precision of the final answers – i.e. there is nothing to lose by applying them in the wrong circumstances, apart from a small increase in computing labour, which is not very serious in these days when matrix inversion on an electronic computer is so simple.

5.8 Use of orthonormal functions

Ermakov and Zolotukhin [12] have described a general method of Monte Carlo integration based on orthonormal functions. Let $\phi_i(\mathbf{y})(i = 0, 1, \ldots, n)$ be a system of functions orthonormal over a region D of the space of vectors $\mathbf{y}$: that is to say

$$\int_D \phi_i \phi_j d\mathbf{y} = \begin{cases} 1 \text{ if } i = j. \\ 0 \text{ if } i \neq j. \end{cases} \qquad (5.8.1)$$

Write $\omega = \omega(\mathbf{y}_0, \mathbf{y}_1, \ldots, \mathbf{y}_n)$ for the $(n+1) \times (n+1)$ determinant having $\phi_i(\mathbf{y}_j)$ as the element in the ith row and jth column $(i, j = 0, 1, \ldots, n)$, and write ω_f for the corresponding determinant in which $f(\mathbf{y}_j)$ replaces $\phi_0(\mathbf{y}_j)$ in the zeroth row $(j = 0, 1, \ldots, n)$. Then the following identity† holds for any pair of functions $f = f(\mathbf{y})$ and $g = g(\mathbf{y})$:

$$\int_D \cdots \int_D \frac{\omega_f \omega_g}{(n+1)!} \, d\mathbf{y}_0 \, d\mathbf{y}_1 \ldots d\mathbf{y}_n =$$

$$= \int_D fg \, d\mathbf{y} - \sum_{i=1}^{n} \left[\int_D f\phi_i \, d\mathbf{y} \right] \left[\int_D g\phi_i \, d\mathbf{y} \right]. \quad (5.8.2)$$

Putting $g = \phi_0$ in (5.8.2), we get

$$\int_D \cdots \int_D \frac{\omega_f}{\omega} \cdot \frac{\omega^2}{(n+1)!} \, d\mathbf{y}_0 \ldots d\mathbf{y}_n = \int_D f\phi_0 \, d\mathbf{y}; \quad (5.8.3)$$

and putting $g = f$ in (5.8.2), we get

$$\int_D \cdots \int_D \left[\frac{\omega_f}{\omega} \right]^2 \cdot \frac{\omega^2}{(n+1)!} \, d\mathbf{y}_0 \ldots d\mathbf{y}_n$$

$$= \int_D f^2 \, d\mathbf{y} - \sum_{i=1}^{n} \left[\int_D f\phi_i \, d\mathbf{y} \right]^2. \quad (5.8.4)$$

Finally putting $f = \phi_0$ in (5.8.4) we have

$$\int_D \cdots \int_D \frac{\omega^2}{(n+1)!} \, d\mathbf{y}_0 \ldots d\mathbf{y}_n = 1. \quad (5.8.5)$$

Equation (5.8.5) shows that $\omega^2/(n+1)!$ is a joint probability density function. Hence if we sample $\eta_0, \eta_1, \ldots, \eta_n$ from this distribution with density function $\omega^2/(n+1)!$, the estimator

$$t = \omega_f(\eta_0, \ldots, \eta_n)/\omega(\eta_0, \ldots, \eta_n) \quad (5.8.6)$$

† This identity ought to be a standard result in the theory of orthonormal functions, but we cannot give a reference to it. It is easily proved by expanding the determinants and using (5.8.1) to remove unwanted terms in the expansion.

is an unbiased estimator of

$$\theta = \int_D f(\mathbf{y}) \, \phi_0(\mathbf{y}) \, d\mathbf{y}, \qquad (5.8.7)$$

by virtue of (5.8.3). Further, by subtracting the square of (5.8.3) from (5.8.4), we have

$$\text{var } t = \int_D f^2 \, dy - \sum_{i=0}^{n} \left[\int_D f \phi_i \, dy \right]^2$$

$$= \inf \int_D \left[f - \sum_{i=0}^{n} c_i \phi_i \right]^2 dy, \qquad (5.8.8)$$

where the infinum is taken over all c_i. Thus, in terms of the usual metric in the orthonormal space, the standard error of t equals the distance of the function f from the subspace spanned by $\phi_0, \ldots, \phi_n$. In particular t has zero variance if f is a linear combination of $\phi_0, \phi_1, \ldots, \phi_n$.

This method offers great possibilities, especially since it may be applied in any number of dimensions, but it depends on the solution of two preliminary problems: first, we must construct $n+1$ functions orthonormal over the region D; second, we must find an efficient way of sampling $\eta_0, \eta_1, \ldots, \eta_n$ with joint probability density function $[\omega(\eta_0, \ldots, \eta_n)]^2/(n+1)!$. Even then the computation of t is in general no small matter. It is however, worth noting that the sampling method depends only upon D and $\phi_0, \ldots, \phi_n$, and that the estimator is a linear functional of f, whose coefficients depend only upon D and $\phi_0, \ldots, \phi_n$. Hence the prospect is perhaps brighter if we have a large number of different functions f to integrate over the same fixed region D, so that we can afford to sink a certain amount of capital into determining a fixed set of $\phi_0, \ldots, \phi_n$, a fixed sampling scheme, and a fixed routine for computing the coefficients in the functional.

Let us consider the application of this method to the estimation of

$$\theta = \int_0^1 g(x) \, dx, \qquad (5.8.9)$$

and let us take $n = 0$, so that we have a single orthonormal function

$\phi(x) = \phi_0(x)$ and the determinant ω is trivial. Our estimator is then

$$t = f(\eta)/\phi(\eta) = g(\eta)/[\phi(\eta)]^2, \qquad (5.8.10)$$

where η is sampled from the distribution with density function $[\phi(y)]^2$ and where $g(x) = f(x)\phi(x)$. If we make the obvious choice, $\phi(x) = 1$, this is simply crude sampling.

We can be more subtle than this, however. Let

$$g^*(x) = g(x) - (1-x)g(0) - xg(1). \qquad (5.8.11)$$

Then $\{g(x) - g^*(x)\}$ is a linear function that we can integrate directly,

$$\int\limits_0^1 \{g(x) - g^*(x)\}\,dx = \tfrac{1}{2}\{g(0) + g(1)\}, \qquad (5.8.12)$$

so that it remains to estimate

$$\theta^* = \int\limits_0^1 g^*(x)\,dx. \qquad (5.8.13)$$

Now we know that $g^*(0) = g^*(1) = 0$, and this knowledge can guide us to a better choice of an orthonormal function. Instead of $\phi(x) = 1$, we take

$$\phi(x) = \{6x(1-x)\}^{1/2}, \qquad (5.8.14)$$

which has the same properties as g^*. This leads us to estimate θ by the formula

$$t = \tfrac{1}{2}\{g(0) + g(1)\} + \frac{1}{6\eta(1-\eta)}\{g(\eta) - (1-\eta)g(0) - \eta g(1)\}, \qquad (5.8.15)$$

where the distribution of η has density function $6y(1-y)$. We may further improve on this by the use of antithetic variates (5.6.2), giving the estimator

$$t = \tfrac{1}{2}\{g(0) + g(1)\} + \frac{1}{12\eta(1-\eta)}\{g(\eta) + g(1-\eta) - g(0) - g(1)\}. \qquad (5.8.16)$$

Formula (5.8.15) is a zero-variance estimator for all quadratic polynomials, and (5.8.16) for all cubic polynomials.

As a further step, let

$$g^*(x) = g(x) - (1-x)(1-2x)g(0) + x(1-2x)g(1) - 4x(1-x)g(\tfrac{1}{2}).$$

$$(5.8.17)$$

Then we can integrate $\{g(x) - g^*(x)\}$ directly, this time by Simpson's rule:

$$\int\limits_0^1 \{g(x) - g^*(x)\}\, dx = \tfrac{1}{6}\{g(0) + g(1) + 4g(\tfrac{1}{2})\}. \qquad (5.8.18)$$

This time we know that $g^*(0) = g^*(1) = g^*(\tfrac{1}{2}) = 0$, and it is convenient to take

$$\phi(x) = \{30x(1-x)\}^{1/2}(1-2x). \qquad (5.8.19)$$

This leads us to the estimator

$$t = \tfrac{1}{6}\{g(0) + g(1) + 4g(\tfrac{1}{2})\} +$$

$$+ \frac{1}{30\eta(1-\eta)(1-2\eta)^2}\{g(\eta) - (1-\eta)(1-2\eta)g(0) +$$

$$+ \eta(1-2\eta)g(1) - 4\eta(1-\eta)g(\tfrac{1}{2})\}, \qquad (5.8.20)$$

where the distribution of η has density function $30y(1-y)(1-2y)^2$. Using antithetic variates as before, we may improve this to

$$t = \tfrac{1}{6}\{g(0) + g(1) + 4g(\tfrac{1}{2})\} + \frac{1}{60\eta(1-\eta)(1-2\eta)^2}\{g(\eta) + g(1-\eta) -$$

$$- (1-2\eta)^2 g(0) - (1-2\eta)^2 g(1) - 8\eta(1-\eta)g(\tfrac{1}{2})\}$$

$$= \tfrac{1}{6}\{g(0) + g(1) + 4g(\tfrac{1}{2})\}$$

$$+ \frac{1}{60\eta(1-\eta)}\{g(\eta) + g(1-\eta) - g(0) - g(1)\} +$$

$$+ \frac{1}{15(1-2\eta)^2}\{g(\eta) + g(1-\eta) - 2g(\tfrac{1}{2})\}. \qquad (5.8.21)$$

Formulae (5.8.20) and (5.8.21) are zero-variance estimators for quartic and quintic polynomials, respectively.

Random variables with the required density functions are fairly easily obtained from rectangularly distributed random numbers. If $\xi_1 \geqslant \xi_2 \geqslant \xi_3$ are rearranged in order of magnitude, then ξ_2 has density function $6y(1-y)$. If ξ_1 to ξ_5 are independent random numbers arranged so that

$$|\xi_1 - \tfrac{1}{2}| \leqslant |\xi_2 - \tfrac{1}{2}| \leqslant \dots \leqslant |\xi_5 - \tfrac{1}{2}|, \qquad (5.8.22)$$

and if
$$\eta = \begin{cases} \xi_4 \text{ with probability } \tfrac{3}{4} \\ \xi_3 \text{ with probability } \tfrac{1}{4}, \end{cases} \qquad (5.8.23)$$

then η has density function $30y(1-y)(1-2y)^2$.

The estimators (5.8.15) and (5.8.16) have infinite variance unless $g(x)$ satisfies certain continuity conditions at $x = 0$ and $x = 1$; a Lipschitz condition of positive order is sufficient. The estimators (5.8.20) and (5.8.21) require an additional condition at $x = \tfrac{1}{2}$; there a Lipschitz condition of order exceeding $\tfrac{1}{2}$ is sufficient.

5.9 Miscellaneous remarks

It should almost go without saying, if it were not so important to stress it, that whenever in the Monte Carlo estimation of a multiple integral we are able to perform part of the integration by analytical means, that part should be so performed. As in some other kinds of gambling, it pays to make use of one's knowledge of form. An example of the advantage of this course of action is well brought out by Cerulus and Hagedorn [13] in the evaluation of multiple phase space integrals for elementary particles in nuclear physics.

Throughout this chapter we have concentrated on estimators which are linear functions of the integrand. Sometimes, however, it is possible to construct other types of estimators. Mantel [14] gives an interesting example of a quadratic estimator in connexion with the Buffon needle estimation of π, which may be compared with the more usual linear estimators [5], [15], and [16].

Mantel notes that, if we throw a straight needle of length L onto a plane, ruled with a square grid consisting of two sets of equally unit-spaced parallel lines, the two sets being at right angles, then n, the number of intersections of the needle with the lines, has expectation

$$\mathscr{E}n = 4L/\pi \qquad (5.9.1)$$

and variance

$$\operatorname{var} n = \left(1 + \frac{2}{\pi} - \frac{16}{\pi^2}\right) L^2. \qquad (5.9.2)$$

The usual procedure, involving linear estimators, equates (5.9.1) to the observed number of intersections. Mantel equates (5.9.2) to the observed variance, and solves the resulting quadratic equation in π. For an experiment with 101 throws and $L = 1$, he obtains a 90% confidence interval for π, which is 3·09 to 3·19 in the case of (5.9.1), and which is 3·138 to 3·146 in the case of (5.9.2).

CHAPTER 6

Conditional Monte Carlo

6.1 Basic formulae

Let α be a random vector distributed over a space $\mathfrak{A}$ with probability density function $f(\alpha)$. If $\mathfrak{A}$ is an awkward space or f is a complicated function, it may be hard to estimate

$$\theta = \mathscr{E}\phi(\alpha). \tag{6.1.1}$$

One way of dealing with this situation is to embed $\mathfrak{A}$ in a product space $\mathfrak{C} = \mathfrak{A} \times \mathfrak{B}$, with a suitably chosen space $\mathfrak{B}$. Each point of $\mathfrak{C} = \mathfrak{A} \times \mathfrak{B}$ can be written in the form $\mathbf{c} = (\mathbf{a}, \mathbf{b})$ where $\mathbf{a}$ and $\mathbf{b}$ are points of $\mathfrak{A}$ and $\mathfrak{B}$ respectively. We can regard $\mathbf{a}$ and $\mathbf{b}$ as the first and second co-ordinates of $\mathbf{c}$; so that $\mathbf{a}$ is a function of $\mathbf{c}$ which maps the points of $\mathfrak{C}$ onto $\mathfrak{A}$. Similarly, if we sample a random vector $\gamma = (\alpha, \beta)$ from $\mathfrak{C}$ with probability density function $h(\mathbf{c})$, we have a mapping of γ to α, which is a random vector of $\mathfrak{A}$. The α obtained in this way will not in general have the desired density f; but we can compensate for this by means of an appropriate weighting function.

Let $d\mathbf{c}$ denote the volume element swept out in $\mathfrak{C}$ when $\mathbf{a}$ and $\mathbf{b}$ sweep out volume elements $d\mathbf{a}$ and $d\mathbf{b}$ in $\mathfrak{A}$ and $\mathfrak{B}$ respectively. The Jacobian of the transformation $\mathbf{c} = (\mathbf{a}, \mathbf{b})$ is accordingly

$$J(\mathbf{c}) = J(\mathbf{a}, \mathbf{b}) = d\mathbf{a}\, d\mathbf{b}/d\mathbf{c}. \tag{6.1.2}$$

Let $g(\mathbf{c}) = g(\mathbf{a}, \mathbf{b})$ be an arbitrary real function defined on $\mathfrak{C}$ such that

$$G(\mathbf{a}) = \int_{\mathfrak{B}} g(\mathbf{a}, \mathbf{b})\, d\mathbf{b} \tag{6.1.3}$$

never vanishes. We shall also suppose that $h(\mathbf{c})$ is never zero. We define the weight function

$$w(\mathbf{c}) = f(\mathbf{a})\, g(\mathbf{c})\, J(\mathbf{c})/G(\mathbf{a})\, h(\mathbf{c}). \tag{6.1.4}$$

Here, of course, **a** is the first co-ordinate of **c**. Then we have the following identity:

$$
\begin{aligned}
\int_{\mathfrak{A}} \phi(\mathbf{a}) f(\mathbf{a})\, d\mathbf{a} &= \int_{\mathfrak{A}} d\mathbf{a}\, \frac{\phi(\mathbf{a}) f(\mathbf{a})}{G(\mathbf{a})} \int_{\mathfrak{B}} g(\mathbf{a}, \mathbf{b})\, d\mathbf{b} \\
&= \int_{\mathfrak{A} \times \mathfrak{B}} \frac{\phi(\mathbf{a}) f(\mathbf{a}) g(\mathbf{c})}{G(\mathbf{a}) h(\mathbf{c})}\, h(\mathbf{c})\, d\mathbf{a}\, d\mathbf{b} \\
&= \int_{\mathfrak{C}} \phi(\mathbf{a})\, w(\mathbf{c})\, h(\mathbf{c})\, \frac{d\mathbf{a}\, d\mathbf{b}}{J(\mathbf{c})} \\
&= \int_{\mathfrak{C}} \phi(\mathbf{a})\, w(\mathbf{c})\, h(\mathbf{c})\, d\mathbf{c}. \qquad (6.1.5)
\end{aligned}
$$

This shows that if α is the first co-ordinate of a random vector γ, sampled from $\mathfrak{C}$ with density function $h(\mathbf{c})$, then

$$
t = \phi(\alpha)\, w(\gamma) \qquad (6.1.6)
$$

is an unbiased estimator of the required θ defined in (6.1.1). Both $\mathfrak{B}$ and h are at our disposal; we may choose them to simplify the sampling procedure. The function g plays the rôle of an importance function and we may select it to minimize variations in t and increase the precision of the estimator.

6.2 Conditional Monte Carlo

Conditional Monte Carlo is a special case of the foregoing theory. In it we start from a given distribution $h(\mathbf{c})$ on the product space $\mathfrak{C} = \mathfrak{A} \times \mathfrak{B}$, and we are told that $f(\mathbf{a}) = f(\mathbf{a}, \mathbf{b}_0)$ is the conditional distribution of $h(\mathbf{c})$ given that $\mathbf{b} = \mathbf{b}_0$. If we write $\psi(\mathbf{b})$ for the probability density function of β when $\gamma = (\alpha, \beta)$ has density function $h(\mathbf{c})$, we have

$$
h(\mathbf{c})\, d\mathbf{c} = f(\mathbf{a}, \mathbf{b})\, \psi(\mathbf{b})\, d\mathbf{a}\, d\mathbf{b}, \qquad (6.2.1)
$$

and comparison of (6.1.2) and (6.2.1) gives

$$
J(\mathbf{c}) = h(\mathbf{c})/f(\mathbf{a}, \mathbf{b})\, \psi(\mathbf{b}). \qquad (6.2.2)
$$

In particular

$$J(\mathbf{a}, \mathbf{b}_0) = h(\mathbf{a}, \mathbf{b}_0)/f(\mathbf{a})\,\psi(\mathbf{b}_0). \tag{6.2.3}$$

By eliminating $f(\mathbf{a})$ from (6.1.4) and (6.2.3) we get

$$w(\mathbf{c}) = \frac{h(\mathbf{a}, \mathbf{b}_0)}{h(\mathbf{a}, \mathbf{b})}\frac{J(\mathbf{a}, \mathbf{b})}{J(\mathbf{a}, \mathbf{b}_0)}\frac{g(\mathbf{a}, \mathbf{b})}{\psi(\mathbf{b}_0)\,G(\mathbf{a})}. \tag{6.2.4}$$

This leads to the following rule: *Suppose that $\gamma = (\alpha, \beta)$ is distributed over $\mathfrak{C}$ with probability density function $h(\mathbf{c}) = h(\mathbf{a}, \mathbf{b})$. Then*

$$t = \phi(\alpha)\,w(\gamma), \tag{6.2.5}$$

where $w(\gamma)$ is given by (6.2.4), *is an unbiased estimator of the conditional expectation of $\phi(\alpha)$ given that $\beta = \mathbf{b}_0$.*

Notice that this rule requires neither sampling from the possibly awkward space $\mathfrak{A}$ nor evaluation of the possibly complicated function f. As before, g is available for variance reduction. The theory provides for several simultaneous conditions, since $\beta = \mathbf{b}_0$ is a vector condition.

6.3 Positive-homogeneous conditions of the first degree

We now consider a special case of the theory in § 6.2. Suppose that $\mathfrak{C}$ is an m-dimensional Euclidean space. We express $\mathfrak{C}$ as a product of spaces $\mathfrak{A}$ and $\mathfrak{B}$, neither of which need be Euclidean. We suppose however that $\mathfrak{B}$ is one-dimensional, so that we may write $\mathbf{c} = (\mathbf{a}, b)$ where $b = b(\mathbf{c})$ is a scalar function of $\mathbf{c}$. We also suppose that $b(\mathbf{c})$ is a positive-homogeneous function of the first degree; that is to say

$$b(\rho\mathbf{c}) = \rho b(\mathbf{c}) \tag{6.3.1}$$

for all $\mathbf{c}$ and for all $\rho > 0$. The equation

$$b(\mathbf{c}) = b_0 \tag{6.3.2}$$

then represents an $(m-1)$-dimensional surface in $\mathfrak{C}$, and, as b_0 varies, we get a family of similarly situated and similarly shaped concentric surfaces with the origin $\mathbf{c} = \mathbf{0}$ as centre. Associated with any point $\mathbf{c}$ there will be a scale factor defined by

$$\lambda = b_0/b(\mathbf{c}). \tag{6.3.3}$$

Let us choose the space $\mathfrak{A}$ to be the space of all directions through the origin in $\mathfrak{C}$: that is to say, when $\mathbf{c} = (\mathbf{a}, b)$, $\mathbf{a}$ denotes the direction of the straight line from the origin to $\mathbf{c}$. It then follows from (6.3.1) and (6.3.3) that

$$\rho\mathbf{c} = (\mathbf{a}, \rho b) = (\mathbf{a}, \rho b_0/\lambda). \qquad (6.3.4)$$

Thus the straight line from the origin to $\mathbf{c}$ cuts the surface (6.3.2) at the point

$$(\mathbf{a}, b_0) = \lambda \mathbf{c}. \qquad (6.3.5)$$

When $\mathbf{a}$ sweeps out an element of solid angle $d\mathbf{a}$ and b sweeps out db, the resulting volume element $d\mathbf{c}$ will be the region of space in the cone $d\mathbf{a}$ between the concentric $(m-1)$-dimensional surfaces b and $b + db$, and will therefore have a volume proportional to b^{m-1}. Hence the Jacobian is

$$J(\mathbf{c}) = d\mathbf{a}\, db/d\mathbf{c} = \Lambda(\mathbf{a})/b^{m-1}, \qquad (6.3.6)$$

where Λ is a function of $\mathbf{a}$ alone. The ratio, occurring in (6.2.4),

$$\frac{J(\mathbf{a}, b)}{J(\mathbf{a}, b_0)} = \left(\frac{b_0}{b}\right)^{m-1} = \lambda^{m-1}. \qquad (6.3.7)$$

We consider the special case in which also $g(\mathbf{a}, b)$ is a function of b and b_0 only, or (what amounts to the same thing) a function of λ and b_0 only. We write

$$g(\mathbf{a}, b) = \delta(\lambda, b_0). \qquad (6.3.8)$$

Since g is independent of $\mathbf{a}$, $G(\mathbf{a})$ will be a constant; by inserting an appropriate constant multiplier in the definition (6.3.8), we can take $G(\mathbf{a}) = 1$. Thus, by (6.1.3),

$$1 = G(\mathbf{a}) = \int_{\mathfrak{B}} g(\mathbf{a}, b)\, db = \int_{\mathfrak{B}} \delta\left(\frac{b_0}{b}, b_0\right) db. \qquad (6.3.9)$$

Insertion of (6.3.5), (6.3.7), (6.3.8) and (6.3.9) into (6.2.4) gives

$$w(\mathbf{c}) = \frac{h(\lambda \mathbf{c})}{h(\mathbf{c})} \lambda^{m-1} \frac{\delta(\lambda, b_0)}{\psi(b_0)}, \qquad (6.3.10)$$

where δ is any function satisfying (6.3.9), and λ is defined by (6.3.3).

The theory in § 6.3 is due to Trotter and Tukey [1], who first discovered conditional Monte Carlo; Hammersley [2] gave the more general formulation of § 6.1 and § 6.2. Wendel [3] examines the theory from a group-theoretic aspect. We now turn to a particular application of (6.3.10), due to Arnold *et al.* [4].

6.4 Practical application

Let $\eta_1, \eta_2, ..., \eta_n$ $(n \geqslant 4)$ be a set of n independent observations from a standardized normal distribution; denote them by $\zeta_1, \zeta_2, ..., \zeta_n$ when arranged in increasing order of magnitude. Define

$$
\begin{aligned}
\beta &= \zeta_n - \zeta_1, \\
\sigma &= \max(\zeta_n - \zeta_2, \zeta_{n-1} - \zeta_1), \\
\tau &= \max(\zeta_n - \zeta_3, \zeta_{n-1} - \zeta_2, \zeta_{n-2} - \zeta_1).
\end{aligned}
\tag{6.4.1}
$$

We are given a rather small number $\epsilon(0 < \epsilon < 1)$ and a pair of numbers S and T such that

$$
P(\sigma \geqslant S \text{ or } \tau \geqslant T) < \epsilon.
\tag{6.4.2}
$$

Here 'or' means 'either or both'. The problem is to find a number B such that

$$
P(\sigma \geqslant S \text{ or } \tau \geqslant T \text{ or } \beta \geqslant B) = \epsilon.
\tag{6.4.3}
$$

The original example [4] had $\epsilon = 0.05$, $S = 3.31$, $T = 3.17$, $n = 4$.

If we were to tackle this problem by direct simulation, we might draw N (say $N = 1000$) sets of 4 independent standardized normal variables, calculate β, σ, τ for each of the N quadruplets, and determine B as the largest value of b such that

$$
\sigma \geqslant 3.31 \text{ or } \tau \geqslant 3.17 \text{ or } \beta \geqslant b
\tag{6.4.4}
$$

held for just $0.05N$ of these quadruplets. But this procedure would be very inaccurate unless N was prohibitively large, for the 5% tail of a sample of 1000 observations may include between 37 and 64 observations, and the values of β corresponding to the 37th and 64th most extreme observations will differ fairly widely. There is also the rather unlikely risk of finding more than 50 quadruplets with $\sigma \geqslant 3.31$ or $\tau \geqslant 3.17$.

To proceed by conditional Monte Carlo write (6.4.3) as

$$Q(B) = \epsilon, \qquad (6.4.5)$$

where

$$Q(b) = \int_T^b P(\sigma \geqslant S \text{ or } \tau \geqslant T | \beta = b_0)\,\psi(b_0)\,db_0 + \int_b^\infty \psi(b_0)\,db_0. \qquad (6.4.6)$$

Here $\psi(b)$ is the probability density function of β, the range of the sample. We can find the second integral in (6.4.6) from tables; so, if we can estimate

$$P(\sigma \geqslant S \text{ or } \tau \geqslant T | \beta = b_0)\,\psi(b_0) \qquad (6.4.7)$$

as a function of b_0, we can evaluate the first integral in (6.4.6) by numerical quadrature and then solve (6.4.5) by inverse interpolation. Accordingly, when $\beta = b_0$, we define

$$\phi(\gamma) = \begin{cases} \psi(b_0) \text{ if } \sigma \geqslant S \text{ or } \tau \geqslant T \\ 0 \text{ otherwise.} \end{cases} \qquad (6.4.8)$$

Here γ is a vector somehow representing the set $\eta_1, \ldots, \eta_n$; we shall give a precise specification of γ in a moment. We see that (6.4.7) is the conditional expectation of (6.4.8) given that $\beta = b_0$.

We now turn to the question of specifying γ and its space $\mathfrak{C}$. Write $\boldsymbol{\eta}$ for the vector with co-ordinates $(\eta_1, \eta_2, \ldots, \eta_n)$ and $\mathfrak{H}$ for n-dimensional Euclidean space. We could take $\gamma = \boldsymbol{\eta}$ and $\mathfrak{C} = \mathfrak{H}$, but we can do better than this, for we notice that β, σ, τ in (6.4.1) are all independent of $\bar{\eta}$, the average of $\eta_1, \ldots, \eta_n$. If we sampled from $\mathfrak{H}$, the errors of our final estimates would be inflated by irrelevant sampling variations in $\bar{\eta}$, and we can avoid this inflation by integrating out $\bar{\eta}$ analytically before commencing the Monte Carlo attack. We define $\mathfrak{C}$ to be the $(n-1)$-dimensional Euclidean subspace of $\mathfrak{H}$ which is orthogonal to the unit vector $\mathbf{u} = (1, \ldots, 1)$ of $\mathfrak{H}$. We then define γ as the projection of $\boldsymbol{\eta}$ on $\mathfrak{C}$. We have

$$\boldsymbol{\eta} = \gamma + \bar{\eta}\mathbf{u}. \qquad (6.4.9)$$

The distribution of $\boldsymbol{\eta}$ is multinormal and, by the orthogonality of γ and $\mathbf{u}$ breaks into two independent parts

$$(2\pi)^{-n/2} e^{-\eta^2/2} = (2\pi)^{-(n-1)/2} e^{-\gamma^2/2} (2\pi)^{-1/2} e^{-\bar{\eta}^2 \mathbf{u}^2/2}, \qquad (6.4.10)$$

where η^2 is the square of the length of η. On integrating out the second term in (6.4.10), we find that the distribution of γ over $\mathfrak{C}$ has density function

$$h(\mathbf{c}) = (2\pi)^{-(n-1)/2} e^{-\mathbf{c}^2/2}, \qquad (6.4.11)$$

where we have now written $\mathbf{c}$ in place of γ. We also have

$$\gamma^2 = (\eta_1 - \bar{\eta})^2 + \ldots + (\eta_n - \bar{\eta})^2. \qquad (6.4.12)$$

We notice that β is a positive-homogeneous function of γ of the first degree, so that we are now ready to apply § 6.3 with $m = n - 1$. From (6.3.10) and (6.4.11) we have

$$w(\mathbf{c}) = e^{(1-\lambda^2)\mathbf{c}^2/2} \lambda^{n-2} \delta(\lambda, b_0)/\psi(b_0). \qquad (6.4.13)$$

In (6.4.8) we defined $\phi(\gamma)$ for $\beta = b_0$. To extend this to the whole space $\mathfrak{C}$, we have to make $\phi(\gamma)$ constant upon straight lines through the origin $\mathbf{c} = \mathbf{0}$, because in § 6.3 we took $\mathfrak{A}$ to be a space whose points were represented by such lines. Since σ and τ are both, like β, positive-homogeneous functions of γ of the first degree, the ratios σ/β and τ/β are constant along a line through the origin. Consequently the appropriate definition of $\phi(\gamma)$ is

$$\phi(\gamma) = \begin{cases} \psi(b_0) \text{ if } \sigma/\beta \geqslant S/b_0 \text{ or } \tau/\beta \geqslant T/b_0 \\ 0 \text{ otherwise.} \end{cases} \qquad (6.4.14)$$

Since this function is either $\psi(b_0)$ or 0, we cannot do much in the way of flattening it by choosing the multiplier δ in (6.4.13), but we can flatten w by an appropriate choice of δ. In the first place $\lambda^{n-2} = (b_0/b)^{n-2}$ and we shall flatten this if δ contains a factor b^{n-2}. Secondly we want to flatten the exponential term $e^{(1-\lambda^2)\mathbf{c}^2/2}$ in w, and we can do this by a term of the form $e^{-v^2 b^2/2}$, where v is a constant to be determined, for $\mathbf{c}^2$ and b^2 are proportional to one another. Hence we try

$$\delta(\lambda, b_0) = \delta\left(\frac{b_0}{b}, b_0\right) = k b^{n-2} e^{-v^2 b^2/2}, \qquad (6.4.15)$$

where k is another constant. We have to satisfy (6.3.9) and $\mathfrak{B}$ is the half-line $0 \leqslant b \leqslant \infty$; so

$$\int_0^\infty k b^{n-2} e^{-v^2 b^2/2} \, db = 1. \qquad (6.4.16)$$

From (6.4.16) we deduce

$$k = v^{n-1}/2^{(n-3)/2}\, \Gamma(\tfrac{1}{2}n - \tfrac{1}{2}). \qquad (6.4.17)$$

Therefore (6.4.13), (6.4.15) and (6.4.17) yield, after a little simplification,

$$w(\mathbf{c}) = [K(b_0)/\psi(b_0)]\, e^{R(\mathbf{c})}, \qquad (6.4.18)$$

where

$$R(\mathbf{c}) = \tfrac{1}{2}(1 - \lambda^2)(\mathbf{c}^2 - v^2 b^2) \qquad (6.4.19)$$

and

$$K(b_0) = \frac{2(vb_0/\sqrt{2})^{n-1}\, e^{-v^2 b_0^2/2}}{b_0\, \Gamma(\tfrac{1}{2}n - \tfrac{1}{2})}. \qquad (6.4.20)$$

Let us now summarize our procedure. We draw a sample of n independent observations $\eta_1, \ldots, \eta_n$ from the standardized normal distribution and arrange them in increasing order of magnitude $\zeta_1, \ldots, \zeta_n$. We calculate β, σ, τ from (6.4.1), and γ^2 from (6.4.12), where $\bar{\eta} = (\eta_1 + \ldots + \eta_n)/n$. Next we calculate

$$\lambda = b_0/\beta \quad \text{and} \quad R(\gamma) = \tfrac{1}{2}(1 - \lambda^2)(\gamma^2 - v^2 \beta^2). \qquad (6.4.21)$$

Finally, with $K(b_0)$ defined by (6.4.20), we calculate

$$t = \begin{cases} K(b_0)\, e^{R(\gamma)} & \text{if } \sigma/\beta \geqslant S/b_0 \text{ or } \tau/\beta \geqslant T/b_0 \\ 0 & \text{otherwise,} \end{cases} \qquad (6.4.22)$$

which is an unbiased estimator of the required integrand (6.4.7). We repeat this procedure N times, using fresh random $\eta_1, \ldots, \eta_n$ each time, and use the mean of the N values of t to estimate (6.4.7).

The constant v is still at our disposal. When $n = 4$, the case considered in [4], it is easy to show that

$$\tfrac{1}{2} \leqslant \gamma^2/\beta^2 \leqslant 1. \qquad (6.4.23)$$

Thus to flatten $R(\gamma)$, a natural choice is $v^2 = \tfrac{3}{4}$. Arnold *et al.* [4] tried $v^2 = 0\cdot60$ and $0\cdot75$, and found empirically that $v^2 = 0\cdot60$ gave less variation. With $T = 3\cdot17$, $S = 3\cdot31$, $n = 4$ and $\epsilon = 0\cdot05$ one can discover from tables that

$$\int_{3\cdot63}^{\infty} \psi(b_0)\, db_0 = 0\cdot05, \qquad (6.4.24)$$

so that B is greater, but not much greater, than 3·63. So $b_0 = 3·2$, 3·4 3·6, and 3·8 seem reasonable trial values. With these values, and with $N = 1000$, the root of (6.4.5) turned out to be $B = 3·684$, the first integral in $Q(B)$ yielding a mean contribution of 0·00447 to ϵ. Five blocks of 200 successive values of t, gave 0·0040, 0·0056, 0·0047, 0·0038, and 0·0043, for the corresponding values of this first integral. The standard error of the mean of these five values is 0·00032. If we had adopted direct simulation, to obtain equal precision we should have required about ν sets of $\eta_1, \ldots, \eta_n$ instead, where

$$0·00032 = \sqrt{[\epsilon(1 - \epsilon)/\nu]}, \quad \epsilon = 0·05, \tag{6.4.25}$$

giving $\nu = 470,000$. Thus, if we allow 3 times as much computation for each set of $\eta_1, \ldots, \eta_n$, the efficiency of the method is about 160 times that of direct simulation.

There is one further point to be mentioned. We have to calculate (6.4.22) for 4 different values of b_0, but we can do this using the same sets of values of $\eta_1, \ldots, \eta_n$ in each case; each of the 4 corresponding mean values of t will be unbiased estimates, and the linear combination of them, used to evaluate the first integral in (6.4.6), will therefore also be unbiased.

Some more refinements of the Monte Carlo procedure are possible with this problem, though we cannot go into them here; they will be found in [4], where the efficiency is multiplied by a further factor of 10.

CHAPTER 7

Solution of Linear Operator Equations

7.1 Simultaneous linear equations

The applications of the Monte Carlo method that we shall be considering in this chapter are all classical problems of numerical analysis, and we strongly recommend that any reader with a problem of this nature should regard the conventional methods of solution as normal, resorting to the Monte Carlo method only when a very rough approximate solution is wanted as a starting point for later work or when the problem is too large or too intricate to be treated in any other fashion.

Curtiss [1], in 1956, compared the theoretical efficiencies of conventional and Monte Carlo methods in computing one component of the solution **x** of the simultaneous equations (in matrix notation)

$$\mathbf{x} = \mathbf{a} + \mathbf{Hx}, \tag{7.1.1}$$

where **H** is an $n \times n$ matrix and **a** is a given vector. Defining the *norm* of the matrix to be

$$||\mathbf{H}|| = \max_i \left(\sum_j |h_{ij}| \right), \tag{7.1.2}$$

and taking the Monte Carlo method to be the first described below, some of his conclusions were the following:

if $||\mathbf{H}|| > 1$, the Monte Carlo method breaks down;

if $||\mathbf{H}|| = 0.9$, the Monte Carlo method is less efficient than a conventional method in finding a solution to 1% accuracy with $n \leqslant 554$, or to 10% accuracy with $n \leqslant 84$;

if $||\mathbf{H}|| = 0.5$, these figures become: (1%) $n \leqslant 151$, (10%) $n \leqslant 20$.
The accuracy here is measured as a percentage of $\max_i |a_i|$. The state
of affairs has changed slightly since this analysis was performed, but

the conclusions remain essentially true, and support our recommendations.

The first Monte Carlo method of solution is based on one proposed by von Neumann and Ulam [2]. Let $\mathbf{P}$ be another $n \times n$ matrix, such that

$$p_{ij} \geqslant 0, \quad \sum_j p_{ij} \leqslant 1, \qquad (7.1.3)$$

and such that $h_{ij} \neq 0$ implies $p_{ij} \neq 0$. Let

$$p_i = 1 - \sum_j p_{ij}, \qquad (7.1.4)$$

and

$$v_{ij} = h_{ij}/p_{ij} \; (p_{ij} \neq 0), \; = 0 \; (p_{ij} = 0). \qquad (7.1.5)$$

(Von Neumann and Ulam originally considered the special case where these conditions could be met with $p_{ij} = h_{ij}$.)

The matrix $\mathbf{P}$ can then describe a terminating *random walk* (or *Markov chain*; see § 9.1 for further details) on the set of states consisting of the integers from 1 to n. If the walk terminates after k steps, passing through the sequence of integers

$$\gamma = (i_0, i_1, \ldots, i_k), \qquad (7.1.6)$$

then the successive states are connected by the *transition probabilities*

$$P(i_{m+1} = j \,|\, i_m = i, k > m) = p_{ij} \qquad (7.1.7)$$

and the *termination probabilities*

$$P(k = m \,|\, i_m = i, k > m-1) = p_i. \qquad (7.1.8)$$

Define

$$V_m(\gamma) = v_{i_0 i_1} v_{i_1 i_2} \ldots v_{i_{m-1} i_m} \quad (m \leqslant k), \qquad (7.1.9)$$

and

$$X(\gamma) = V_k(\gamma) \, a_{i_k}/p_{i_k}. \qquad (7.1.10)$$

Then the expectation of $X(\gamma)$, conditional on $i_0 = i$, is

$$\sum_\gamma P(\gamma) \, X(\gamma)$$

(where the summation is restricted to those γ with $i_0 = i$)

$$= \sum_{k=0}^{\infty} \sum_{i_1} \cdots \sum_{i_k} p_{ii_1} \cdots p_{i_{k-1}i_k} p_{i_k} v_{ii_1} \cdots v_{i_{k-1}i_k} a_{i_k}/p_{i_k}$$

$$= \sum_{k=0}^{\infty} \sum_{i_1} \cdots \sum_{i_k} h_{ii_1} \cdots h_{i_{k-1}i_k} a_{i_k}$$

$$= \mathbf{a}_i + (\mathbf{Ha})_i + (\mathbf{H}^2\mathbf{a})_i + \ldots \tag{7.1.11}$$

Therefore, provided that the Neumann series $\mathbf{H} + \mathbf{H}^2 + \ldots$ converges (as it will if $||\mathbf{H}|| < 1$), the vector $\mathbf{x}$ with components

$$x_i = \mathscr{E}\{X(\gamma)| i_0 = i\} \tag{7.1.12}$$

is the solution of (7.1.1). We may therefore estimate any particular component x_i by generating walks γ starting out from i and scoring $X(\gamma)$ for each walk when it terminates.

Wasow [3] modifies this scheme by scoring

$$X^*(\gamma) = \sum_{m=0}^{k} V_m(\gamma) a_{i_m}, \tag{7.1.13}$$

which is also an unbiased estimator of x_i. In the special case where $a_i = 1$ $(i = j)$, $= 0(i \neq j$, for some specified $j)$, he shows that the variance of X^* is smaller than that of X if and only if

$$p_j \leqslant \frac{\nu_j}{2 - \nu_j}, \tag{7.1.14}$$

where ν_j is the probability that a walk starting from j never revisits j.

There is also an adjoint method of solving (7.1.1). Let $\mathbf{Q}$ be another $n \times n$ matrix, with

$$q_{ij} \geqslant 0, \quad \sum_j q_{ij} \leqslant 1, \tag{7.1.15}$$

such that $h_{ji} \neq 0$ implies $q_{ij} \neq 0$, and let

$$q_i = 1 - \sum q_{ij} \tag{7.1.16}$$

and

$$w_{ij} = h_{ji}/q_{ij} \ (q_{ij} \neq 0), \ = 0 \ (q_{ij} = 0). \tag{7.1.17}$$

This differs from (7.1.5) in the reversal of the suffices to h. The matrix $\mathbf{Q}$ then describes another Markov chain, in the same way as the matrix $\mathbf{P}$. Let.

$$\pi_i \geqslant 0, \quad \sum \pi_i = 1, \quad \alpha_i = a_i/\pi_i, \tag{7.1.18}$$

and

$$W_m(\gamma) = \alpha_{i_0} w_{i_0 i_1} w_{i_1 i_2} \ldots w_{i_{m-1} i_m} \tag{7.1.19}$$

Then, if the starting point of γ is chosen from the distribution

$$P(i_0 = i) = \pi_i, \tag{7.1.20}$$

we have two estimators of x_j corresponding to (7.1.10) and (7.1.13), for

$$\begin{aligned} x_j &= \mathscr{E}\{W_k(\gamma)\,\delta_{i_k j}/q_{i_k}\} \\ &= \mathscr{E}\left\{\sum_{m=0}^{k} W_m(\gamma)\,\delta_{i_m j}\right\}, \end{aligned} \tag{7.1.21}$$

where δ_{ij} is the Kronecker symbol ($= 1$ or 0 according as $i = j$ or $i \neq j$).

The adjoint method is more suitable for finding the general shape of $\mathbf{x}$ than the direct method, which concentrates on a single element x_i, since the same walk yields estimates of all x_j simultaneously.

7.2 Sequential Monte Carlo

Halton [4] has studied a method of accelerating the preceding processes. After a fair amount of work, one should have a rough estimate $\hat{\mathbf{x}}$ for $\mathbf{x}$. If we put

$$\mathbf{y} = \mathbf{x} - \hat{\mathbf{x}} \tag{7.2.1}$$

and

$$\mathbf{d} = \mathbf{a} + \mathbf{H}\hat{\mathbf{x}} - \hat{\mathbf{x}}, \tag{7.2.2}$$

equation (7.1.1) is transformed into another of similar form,

$$\mathbf{y} = \mathbf{d} + \mathbf{H}\mathbf{y}, \tag{7.2.3}$$

where the elements of $\mathbf{d}$ are considerably smaller than those of $\mathbf{a}$. Consequently we can solve (7.2.3) to the same absolute accuracy as (7.1.1) in a much shorter time. The solution of (7.1.1) is then derived immediately by (7.2.1).

Halton takes this principle further, by improving **x** at regular intervals, and shows that in this case the variance after the sth stage of improvement behaves at least as well as

$$\left(\frac{\sigma}{1-\sigma}\right)^{s-1},$$

where

$$\sigma = \max_i \left\{ \sum_j h_{ij}^2 / p_{ij} \right\},$$

and therefore decreases rapidly if $\sigma < \frac{1}{2}$. This condition can be satisfied by a suitable choice of **P** only if $||\mathbf{H}|| < \frac{1}{2}$.

7.3 Fredholm integral equations of the second kind

The equation

$$f(x) = g(x) + \int K(x,y)f(y)\,dy \qquad (7.3.1)$$

may be solved by similar methods to (7.1.1). There are three possible approaches.

(i) The integral may be replaced by a quadrature formula. The equation then becomes a finite matrix equation like (7.1.1) for the values of f at the quadrature points.

(ii) If we have a complete set of functions ϕ_0, ϕ_1, ..., in terms of which we know the expansion of g,

$$g(x) = a_0\,\phi_0(x) + a_1\,\phi_1(x) + \ldots, \qquad (7.3.2)$$

and for which we know the expansions

$$\int K(x,y)\,\phi_j(y)\,dy = h_{0j}\phi_0(x) + h_{1j}\phi_1(x) + \ldots, \qquad (7.3.3)$$

then, setting

$$f(x) = x_0\,\phi_0(x) + x_1\,\phi_1(x) + \ldots, \qquad (7.3.4)$$

we get the coefficients of the expansion of f by solving the infinite matrix equation

$$\mathbf{x} = \mathbf{a} + \mathbf{Hx}, \qquad (7.3.5)$$

which we solve no differently from the finite equation (7.1.1), apart from taking precautions in case the walk should not terminate.

(iii) We may solve the equation as it stands, by processes analogous to those of § 7.1, transferred from discrete to continuous distributions. For example, by analogy with (7.1.3) onwards, we could define a function p such that

$$p(x, y) \geqslant 0, \quad \int p(x, y)\, dy \leqslant 1 \qquad (7.3.6)$$

and

$$\left. \begin{aligned} p(x) &= 1 - \int p(x, y)\, dy, \quad v(x, y) = K(x, y)/p(x, y), \\ \gamma &= (x_0, x_1, \ldots, x_k), \quad V_m(\gamma) = v(x_0, x_1)\ldots v(x_{m-1}\, x_m), \\ X(\gamma) &= V_k(\gamma)\, g(x_k)/p(x_k). \end{aligned} \right\} \qquad (7.3.7)$$

Then

$$f(x) = \mathscr{E}\{X(\gamma) | x_0 = x\}, \qquad (7.3.8)$$

provided that the series

$$K(x, y) + \int K(x, x_1)\, K(x_1, y)\, dx_1 +$$

$$+ \int\int K(x, x_1)\, K(x_1, x_2)\, K(x_2, y)\, dx_1\, dx_2 + \ldots \qquad (7.3.9)$$

converges. A condition for convergence is that

$$\|K\| = \sup_x \int |K(x, y)|\, dy < 1. \qquad (7.3.10)$$

For further discussion of this method we refer the reader to Cutkosky [5] and Page [6]. A frequently-occurring special case, where $K(x, y) = K(x - y)$ and the integration is over a finite line-segment, is treated graphically in [7].

Since we now have integrals to deal with, it should be possible to use some of the techniques of Chapter 5 to increase efficiency, or one of the quasirandom schemes of § 3.4.

As in the case of simultaneous equations, there are an adjoint method and modifications analogous to those of Wasow and Halton. The analogue of the transformation (7.2.1), (7.2.2) appears in a paper by Albert [8].

7.4 The Dirichlet problem

One of the earliest [9] and most popular illustrations of the Monte Carlo method, because it is so easy to grasp, is the solution of Dirichlet's problem in potential theory. This is unfortunate, since anyone who puts the method to a practical test will soon find it to be very laborious and inefficient, compared to relaxation methods, say, and may be tempted to attach this stigma to Monte Carlo methods as a whole. Nevertheless, this book would be incomplete without some discussion of this problem. Curtiss [10] discusses differential and difference equations in general at greater length.

Dirichlet's problem is to find a function u, defined, continuous, and differentiable over a closed domain D with boundary C, satisfying

$$\nabla^2 u = 0 \text{ on } D, \quad u = f \text{ on } C, \tag{7.4.1}$$

where f is some prescribed function, and ∇^2 is the Laplacian operator. One usually starts by covering D by a cubic mesh, and replacing ∇^2 by its finite-difference approximation. Taking the two-dimensional case for convenience, this approximation is

$$h^{-2}\{u(x, y+h) + u(x, y-h) + u(x+h, y) + u(x-h, y) - 4u(x, y)\}, \tag{7.4.2}$$

where h is the mesh size. In other words the equation $\nabla^2 u = 0$ is replaced by

$$u(x, y) = \tfrac{1}{4}\{u(x, y+h) + u(x, y-h) + u(x+h, y) + u(x-h, y)\}. \tag{7.4.3}$$

Suppose for simplicity that the boundary C lies on the mesh, and consider a random walk (*Pólya walk*) that starts from a given interior point P of D and proceeds by stepping to one of the four neighbouring points at random (the four possible neighbours have equal and independent probabilities at each step) until finally it hits the boundary at a point Q. Then $f(Q)$ is an unbiased estimator of $u(P)$.

To show this, it is sufficient to reduce the problem to the form (7.1.1) where the order of $\mathbf{H}$ is equal to the number of mesh points in D, and where $\mathbf{H}$ has four elements equal to $\tfrac{1}{4}$ in each row corresponding to an interior point of D, all other elements being zero. The random walk is then identical with the first (and, indeed, the second) method of §7.1.

Once again there is an adjoint method, which this time means starting walks out from the boundary. If the starting point is chosen on the boundary with a probability distribution $p(Q)$, and a walk passes through the point P just $n(P)$ times before hitting the boundary again, then an unbiased estimator of $u(P)$ is $\frac{1}{2}n(P)f(Q)/p(Q)$.

The sequential method (§ 7.2) may again be profitably applied.

It can be seen that when we express the problem in matrix form we have $||\mathbf{H}|| = 1$. In view of Curtiss' analysis [1], therefore, it is hardly surprising that the methods turns out to be generally inefficient.

Muller [11] has proposed a method that does not resort to the difference approximation. Let $S(P)$ be the largest sphere (circle) with centre P that does not go outside D. Then it is well known from potential theory that $u(P)$ is equal to the average value of u over the surface of $S(P)$. Muller's method is to take a sequence of points $P_0 = P$, P_1, P_2, ..., where P_{m+1} is a random point on the surface of $S(P_m)$, until a point is reached that is sufficiently near the boundary C. If Q is now the nearest point actually on C, $f(Q)$ is taken for the estimator of $u(P)$. This method may be generalized by taking $S(P)$ to be some other shape than a sphere, provided that one knows the Green's function for that surface. The essence of the procedure is to economize in the number of steps by reaching the boundary of the sphere in one leap instead of via individual steps on an 'equivalent' lattice.

7.5 Eigenvalue problems

The other type of problem connected with linear operators is the eigenvalue problem, of determining values of λ for which the matrix equation

$$\mathbf{Hx} = \lambda\mathbf{x}, \qquad (7.5.1)$$

the integral equation

$$\int K(x,y)f(y)\,dy = \lambda f(x), \qquad (7.5.2)$$

or the differential equation

$$\nabla^2 u(x) - (V(x) - \lambda)\,u(x) = 0, \qquad (7.5.3)$$

to give three typical instances, has a non-trivial solution ($\mathbf{x}$, f, or u)

satisfying certain general conditions. Characteristically, the Monte Carlo method will not find more than the extreme eigenvalues of these equations, and then only under certain conditions on $\mathbf{H}$, K, or V.

The solution of (7.5.1) or (7.5.2) depends upon the convergence, for almost all choices of $\mathbf{x}_0$ or f_0, of the sequence

$$\mathbf{x}_m = \mathbf{H}\mathbf{x}_{m-1}/\lambda_m \tag{7.5.4}$$

or

$$f_m(x) = \int K(x,y) f_{m-1}(y)\,dy/\lambda_m, \tag{7.5.5}$$

where λ_m is chosen to make

$$||\mathbf{x}_m|| = 1 \quad \text{or} \quad ||f_m|| = 1, \tag{7.5.6}$$

where $||\,.\,||$ denotes some vector norm; for instance we may take

$$||\mathbf{x}|| = \sum_i |x_i|, \quad ||f|| = \int |f(x)|\,dx. \tag{7.5.7}$$

Then λ_m converges to the dominant (largest) eigenvalue λ of $\mathbf{H}$ or K, and $\mathbf{x}_m$ converges to the corresponding eigenvector or f_m to the corresponding eigenfunction. We suppose that λ is simple, real, and positive.

Taking the matrix formulation (7.5.1) for convenience, let $\mathbf{Q}$ now be a *stochastic* matrix, i.e. one such that

$$q_{ij} \geqslant 0, \quad \sum_j q_{ij} = 1, \tag{7.5.8}$$

and define w_{ij} by (7.1.17) as before. The Markov chain described by $\mathbf{Q}$ now *never* terminates, giving rise to an infinite sequence of states

$$\gamma = (i_0, i_1, i_2, \ldots). \tag{7.5.9}$$

We define $W_m(\gamma)$ by (7.1.19). Then

$$\mathscr{E}\{W_m(\gamma)\,\delta_{i_m j}|i_0 = i\} = (\mathbf{H}^m)_{ji}\,\alpha_i. \tag{7.5.10}$$

If i_0 is sampled from the distribution (7.1.20), therefore, we have

$$\mathscr{E}\{W_m(\gamma)\,\delta_{i_m j}\} = (\mathbf{H}^m \mathbf{a})_j = \lambda_1 \lambda_2 \ldots \lambda_m(\mathbf{x}_m)_j, \tag{7.5.11}$$

by (7.5.4), taking $\mathbf{x}_0 = \mathbf{a}$; also, summing over j,

$$\mathscr{E}\{W_m(\gamma)\} = \lambda_1 \lambda_2 \ldots \lambda_m \sum_j (\mathbf{x}_m)_j. \tag{7.5.12}$$

If the dominant eigenvector $\mathbf{x}$ of $\mathbf{H}$ has all its elements of the same sign, then $\mathbf{x}_m$ tends to $\mathbf{x}$ as m tends to infinity, and (by (7.5.6), (7.5.7)) we have $\sum\limits_j (\mathbf{x}_m)_j = 1$, so that

$$\mathscr{E}\{W_m(\gamma)\} = \lambda_1\lambda_2\ldots\lambda_m. \qquad (7.5.13)$$

Thus

$$\mathscr{E}\{W_m(\gamma)\}/\mathscr{E}\{W_n(\gamma)\} = \lambda_{n+1}\lambda_{n+2}\ldots\lambda_m \sim \lambda^{m-n} \qquad (7.5.14)$$

as m and n tend to infinity. We may therefore use

$$\hat{\lambda} = [W_m(\gamma)/W_n(\gamma)]^{1/(m-n)} \qquad (7.5.15)$$

as a (biased) estimator of λ.

Having estimated λ by $\hat{\lambda}$, we may estimate the (unnormalized) elements x_j of $\mathbf{x}$ by a formula such as

$$\hat{x}_j = \sum_{r=n+1}^{m} W_r(\gamma)\,\delta_{i_r j}\hat{\lambda}^{m-n-r}(1-\hat{\lambda})/(1-\hat{\lambda}^{m-n}). \qquad (7.5.16)$$

The convergence of (7.5.4), and therefore of the Monte Carlo process, is most rapid if $\mathbf{x}_0$ is close to $\mathbf{x}$. We may be able to approach this situation after performing our sampling by replacing α_{i_0} by $\hat{x}_{i_0}/\pi_{i_0}$ in (7.1.19), thereby giving new weights to the same set of paths γ, which we hope would yield better estimates of λ and $\mathbf{x}$.

The whole of this method applies equally, under suitable conditions, to the integral equation (7.5.2). See [12], for instance.

Fortet [13] proposes a fundamentally different method. Suppose that we have to solve (7.5.2) knowing that the kernel $K(x,y)$ is symmetric and positive definite. Then ([14] § 34) there is a Gaussian process $X(x)$ having K as its covariance function; i.e. $X(x)$ has a normal distribution with variance $K(x,x)$, and $\text{cov}(X(x), X(y)) = K(x,y)$. We may often represent such a Gaussian process sufficiently accurately by a function made up of straight-line segments linking the positions $X(x_i)$, $X(x_{i+1})$ of suitably correlated Gaussian variables at prescribed discrete values x_i. This is equivalent to truncating the expansion of the process in terms of a Schauder basis [15].

Define a random variable

$$Y = \int [X(x)]^2\,dx. \qquad (7.5.17)$$

This has a characteristic function

$$\phi(v) = \mathscr{E} \exp(iv Y) = D(2iv)^{-1/2}, \qquad (7.5.18)$$

where $D(\mu)$ is the Fredholm determinant of the integral equation, whose roots are the inverses of its eigenvalues. Fortet's method is to generate values of sample functions X, so that Y can be derived by numerical quadrature, and to use the values of Y to deduce the form of ϕ near the origin, and hence to find λ. Fortet also gives bounds for the errors in the solution, but an unpublished paper by Cohen and Kac suggests that these bounds are wasteful by a factor of as much as 5000.

Fortet's method is connected with one proposed by Donsker and Kac [16] for finding the smallest eigenvalue of Schrödinger's equation (7.5.3). If $X(x)$ is a *Wiener process* (or *Brownian motion*: see [14] or Doob [17]), i.e. a Gaussian process with covariance function

$$\operatorname{cov}\{X(x), X(y)\} = \min(x, y), (x, y \geqslant 0) \qquad (7.5.19)$$

and if

$$L(t) = \int_{0}^{t} V[X(\tau)] d\tau, \qquad (7.5.20)$$

and

$$Z(s, t) = e^{-sL(t)}, \qquad (7.5.21)$$

then

$$\mathscr{E} Z(1, t) = \sum_{j} e^{-\lambda_j t} \psi_j(0) \int_{-\infty}^{\infty} \psi_j(x) \, dx$$

$$\sim e^{-\lambda_1 t} \psi_1(0) \int_{-\infty}^{\infty} \psi_1(x) \, dx \text{ for large } t. \qquad (7.5.22)$$

Here λ_j denote the eigenvalues in order of increasing magnitude and ψ_j the corresponding eigenfunctions of the equation

$$\frac{1}{2} \frac{d^2}{dx^2} \psi(x) - V(x) \psi(x) + \lambda \psi(x) = 0. \qquad (7.5.23)$$

An estimator of λ_1 is therefore

$$-t^{-1}\log Z(1,t) \text{ for large } t, \tag{7.5.24}$$

or, better,

$$-(t_1-t_2)^{-1}\log\left[Z(1,t_1)/Z(1,t_2)\right] \text{ for large } t_1 \text{ and } t_2. \tag{7.5.25}$$

Kac [18] shows that $Z(1,t)$ may validly be approximated by

$$Y_n(t) = e^{-L_n(t)} \tag{7.5.26}$$

where

$$L_n(t) = \frac{1}{n}\sum_{k<nt} V(n^{-1/2}S_k) \tag{7.5.27}$$

and S_k is the sum of k independent and identically-distributed random variables with zero mean and unit variance.

Equation (7.5.23) is essentially the one-dimensional version of (7.5.3), and Donsker and Kac's method extends to more than one dimension by means of multi-dimensional Wiener processes. A similar approach may be used to derive statistical-mechanical properties, which depend on the distribution of the eigenvalues as a whole (see [19] Chapter 4). There is much scope for further research.

Wasow [20] deals with random walk solutions of the difference approximation to (7.5.3).

For further reading on the general topic of Monte Carlo solution of linear operator problems, see the supplementary references.

CHAPTER 8

Radiation Shielding and Reactor Criticality

8.1 Introduction

We turn now to problems of a more physical kind, beginning with the field in which Monte Carlo methods were first used on a large scale and systematically, namely the flux of uncharged particles through a medium. The point of considering uncharged particles is that their paths between collisions are straight lines, and that they do not influence one another. The latter consideration allows us to take the behaviour of a relatively small sample of particles to represent that of the whole.

There is an element of randomness in these problems from the beginning, and, while it is possible to reduce them to the solution of large integro-differential equations in six dimensions [1] it is most convenient to derive the Monte Carlo methods directly from the physical processes.

Consider a particle (photon or neutron) with energy E, instantaneously at the point $\mathbf{r}$, and travelling in the direction of the unit vector $\boldsymbol{\omega}$. So long as the particle does not collide with an atom of the medium, it will continue to travel in this same direction $\boldsymbol{\omega}$ with this same energy E. However, at each point of its straight path it has a chance of colliding with an atom of the medium. With the usual assumption that the atoms of the surrounding medium are distributed randomly in space, there is a probability $\sigma_c \delta s$ that the particle will collide with an atom while traversing a small length δs of its straight path. The factor of proportionality σ_c is called the *cross-section*: it depends upon the energy E of the particle and also upon the nature of the surrounding medium. If the medium is continuously variable in its constitution, then σ_c will be a variable function of $\mathbf{r}$. But it is

97

more usual to have to deal with situations in which the character of the medium remains homogeneous within each of a small number of distinct regions; over each such region, σ_c will be constant, though σ_c will change abruptly on passing from one region to the next. This sort of situation arises, for example, with uranium rods immersed in water; σ_c is one function of E in the rods and another function of E in the water. It follows that the cumulative distribution function of the distances that the particle travels before collision is

$$F_c(s) = 1 - \exp(-\sigma_c s), \qquad (8.1.1)$$

provided that all points of the path from $\mathbf{r}$ to $\mathbf{r} + s\boldsymbol{\omega}$ are in the same region of the medium.

When a collision takes place, one of three things may happen. The first possibility is that the particle is *absorbed* into the medium: in this case the particle travels no further. The second possibility is that the particle is *scattered*, that is to say it leaves the point of collision in a new direction and with a new energy. The third possibility is *fission* of the struck atom: in this case (which only arises when the original particle is a neutron) several other neutrons, known as *secondary* neutrons, leave the point of collision with various different energies and directions. Each of these possibilities has a certain probability; and conditional probability distributions also govern the second and third possibilities. For instance, given that the third possibility occurs, there is a conditional joint probability distribution for the number of secondary neutrons and for the energies and directions of the emergent particles. These probabilities and distributions are specified by the physics of the problem: for Monte Carlo purposes they are known distributions.

The two problems we shall be concerned with are:

(i) *The Shielding Problem.* When a thick shield of absorbing material is exposed to γ-radiation (photons), of specified energy and angle of incidence, what is the intensity and energy-distribution of the radiation that penetrates the shield?

(ii) *The Criticality Problem.* When a pulse of neutrons is injected into a reactor assembly, will it cause a multiplying chain reaction or will it be absorbed, and, in particular, what is the size of the assembly at which the reaction is just able to sustain itself? This is an eigenvalue problem.

8.2 The elementary approach, and some improvements

A Monte Carlo solution involves the tracking of simulated particles from collision to collision. Starting with a particle, whose energy, direction and position are (E, ω, r), we generate a number s with the exponential distribution (8.1.1) where $\sigma_c = \sigma_c(E)$. If the straight-line path from r to $(r + s\omega)$ does not intersect any boundary (between regions), the particle has a collision at the latter point. Otherwise we allow the particle to proceed as far as the first boundary. If this is the outer boundary, the particle escapes from the system ($\sigma_c = 0$ in a region of free space); if not, we repeat the above procedure replacing r by the boundary point and replacing σ_c by that appropriate to the new region that the particle is entering. The justification for repeating the procedure whenever a boundary is reached lies in the Markovian character of distribution (8.1.1): see § 9.1 for a discussion of Markov processes. (The time of each event may be calculated from the velocity, which depends only on E.)

We shall not go into details of the simulation of the collisions themselves, since the considerations here are physical rather than mathematical; it suffices to say that the result will be a collection (possibly empty) of such particles leaving the collision-point with various energies and directions, that we determine these energies and direction by sampling from the relevant conditional distributions mentioned in § 8.1, and that we follow each of the emergent particles in the same manner as the first, going on in the same fashion as far as the problem requires or time allows.

The above process gives an exact realization of the physical model, but it is not always the most convenient for our purposes, nor the most efficient. For instance, if we are studying a reactor containing a very fissile component, then every neutron entering this region may give rise to a very large number coming out, giving us more tracks than we have time to follow. Here we may turn to the fractional sampling method to which Kahn [2] gives the name 'Russian Roulette'. When the number of particles gets too large, we pick out one of them, and with some probability p we discard it from the sample; otherwise we allow it to continue but multiply its weight (initially unity) by $(1-p)^{-1}$, and we repeat this (with the same or different values of p) until the number of particles is reduced to manageable size.

Conversely, to increase the sample size without introducing a bias, Kahn has the 'splitting' technique [2] [3] where a particle of weight w may be replaced by any number k of identical particles of weights $w_1, \ldots, w_k$, where $w_1 + \ldots + w_k = w$. These particles then proceed independently. One may also avoid losing tracks through absorption; if the absorption probability is α (i.e. α is the conditional probability that absorption occurs, given that a collision has occurred) one replaces σ_c by $\sigma_c(1-\alpha)$ in (8.1.1), and allows only scattering or fission to take place with appropriate relative probabilities. The weight is then multiplied by $\exp(-\sigma_c \alpha s)$ for every segment of path of length s that is traversed in each medium.

By weighting methods such as these, one may control not only the total number of tracks, but also the relative numbers in various regions of space or ranges of energy, thus providing a form of importance sampling. Extending the argument of § 5.4, we should arrange for the number of paths in any class to be proportional to the contribution of that class to the final result, hence, in particular, our avoidance of paths that are absorbed and contribute nothing.

However, we have not yet considered how these samples are to be scored. This is better dealt with under separate headings for the two problems that we are concerned with.

8.3 Special methods for the shielding problem

The outstanding feature of the shielding problem is that, if the shield is worthy of the name, the proportion of photons that penetrate the shield is very small, say one in 10^6. Suppose that we attempt to estimate this proportion by simulating the process and simply counting the photons that emerge. The distribution of the number of emergent photons will be binomial, so that the standard deviation of the proportion of survivors from an original sample of N will be of the order of $10^{-3}N^{-1/2}$. Thus to estimate this proportion to an accuracy of 10% we require the impossible number of 10^8 paths.

Now suppose that the absorption probability is uniform throughout the medium, and that we use the method of the previous section, so that the paths which traverse a distance s in the shield emerge with a weight of $\exp(-\sigma_c \alpha s)$. The error now depends on the variation of s. If the probability ζ of scattering is small, then the paths tend to go straight through the shield, s tends to remain close to the thickness of the shield, and the estimator can be very precise. If, on the other hand, particles are scattered one or more times, then s may vary considerably, and its exponential even more.

Berger and Doggett [4], [5] overcome this difficulty by a semi-analytic method, which incidentally allows the same random paths to be used for shields of other thicknesses. We now suppose always that the shield consists of a uniform infinite slab of thickness t. This simplification of the geometry means that we need work in terms of three co-ordinates only: the energy E, the angle θ between the direction of motion and the normal to the slab (into the slab from the incident face), and the distance z from the incident face of the slab.

In the semi-analytic method, we first generate a random history

$$h = h_n = \left\{ \begin{array}{l} E_0, E_1, \ldots, E_n \\ \theta_0, \theta_1, \ldots, \theta_n \end{array} \right\} \tag{8.3.1}$$

for a particle which undergoes a suitably large number n of scatterings in the medium. Here E_i and θ_i represent the energy and direction immediately after the ith scattering. E_0 and θ_0 are the incident energy and direction. Provided that the particle remains in the medium and is not absorbed before the nth stage, the distribution of h_n is independent of the distances travelled between collisions, and the same is true of any truncation h_r of h_n.

We now define $P_i(\zeta) = P_i(\zeta, t, h)$ to be the probability that a particle has a history h_i (the ith truncation of h) and also crosses the plane $z = \zeta$ between its ith and $(i+1)$th scatterings; and we proceed to write down some recurrence relations for these P_i. In doing so, the abbreviations

$$c_i = \cos\theta_i; \quad \sigma_i = \sigma_c(E_i); \quad \tau_i = [1 - \alpha(E_i)]\sigma_c(E_i) \tag{8.3.2}$$

are convenient. In the first place, it is clear that a straight track with direction θ_i has length $|\varDelta/c_i|$ if it is terminated by two planes with co-ordinates $z = z_0$ and $z = z_0 + \varDelta$. Thus the probability $P_0(\zeta)$ that the particle will pass through $z = \zeta$ before suffering any scatterings is

$$P_0(\zeta) = \exp(-\sigma_0\,\zeta/c_0). \tag{8.3.3}$$

Next consider $P_{i+1}(\zeta)$ for $i \geqslant 0$. This concerns a particle crossing $z = \zeta$ between its $(i+1)$th and $(i+2)$th scatterings. The $(i+1)$th scattering (which has actually been a scattering, since $i \geqslant 0$), may have occurred to the left ($z \leqslant \zeta$) or the right ($z > \zeta$) of the plane $z = \zeta$; suppose it was the left. Then it occurred on some plane $z = \zeta'$ where $0 < \zeta' < \zeta$. After the $(i+1)$th scattering the particle travelled to the right (in order to intersect $z = \zeta$) with energy E_{i+1} and direction θ_{i+1}. Thus $c_{i+1} > 0$. For all this to occur, we have a compound event: (i) that immediately prior to the $(i+1)$th scattering the particle crossed $z = \zeta'$; (ii) that then, travelling with energy E_i and direction θ_i immediately prior to the $(i+1)$th scattering, the particle suffered the $(i+1)$th scattering between the planes $z = \zeta'$ and $z = \zeta' + d\zeta'$; and (iii) that thereupon the particle, now travelling with energy E_{i+1} in direction θ_{i+1}, traversed the region from $z = \zeta'$ to $z = \zeta$ without further collision (either scattering or absorption). The respective probabilities of these events are

$$P(\text{i}) = P_i(\zeta') \tag{8.3.4}$$

$$P(\text{ii}) = \tau_i d\zeta'/|c_i| \tag{8.3.5}$$

and

$$P(\text{iii}) = \exp\{-\sigma_{i+1}(\zeta - \zeta')/c_{i+1}\}. \tag{8.3.6}$$

The Markovian character of the process allows us to combine these three probabilities by multiplying them together. All values of ζ' between 0 and ζ may occur; so

$$P_{i+1}(\zeta) = \int\limits_0^\zeta P_i(\zeta') \exp\{-\sigma_{i+1}(\zeta - \zeta')/c_{i+1}\}\frac{\tau_i d\zeta'}{|c_i|}, \quad c_{i+1} > 0. \tag{8.3.7}$$

When the $(i+1)$th scattering occurs to the right of $z = \zeta$, we have an exactly similar analysis for $c_{i+1} < 0$; and the inequality $\zeta < \zeta' < t$ leads to

$$P_{i+1}(\zeta) = \int\limits_{\zeta}^{t} P_i(\zeta') \exp\{-\sigma_{i+1}(\zeta - \zeta')/c_{i+1}\} \frac{\tau_i d\zeta'}{|c_i|}, \quad c_{i+1} < 0.$$

$$(8.3.8)$$

Equations (8.3.3), (8.3.7) and (8.3.8) have the analytical solution

$$P_i(\zeta) = |c_i| \sum_{j=0}^{i} A_{ij} \exp\{-\sigma_j(\zeta - u_j)/c_j\}, \quad (8.3.9)$$

where u_j denotes 0 or t according as $c_j > 0$ or $c_j < 0$, and where the constants A_{ij} can be successively calculated from the recurrence relations

$$A_{00} = 1/c_0, \quad (8.3.10)$$

$$A_{i+1,j} = \tau_i A_{ij} c_j/(\sigma_{i+1} c_j - \sigma_j c_{i+1}), \quad (8.3.11)$$

$$A_{i+1,i+1} = -\sum_{j=0}^{i} A_{i+1,j} \exp\{-\sigma_j(u_{i+1} - u_j)/c_j\}. \quad (8.3.12)$$

The probability of penetrating the shield is

$$\mathscr{E} \sum_{i=0}^{\infty} P_i(t), \quad (8.3.13)$$

where the expectation $\mathscr{E}$ is taken over all histories h_n for $n = 1, 2, \ldots$ An adequate approximation replaces $\sum_{i=0}^{\infty}$ in (8.3.13) by $\sum_{i=0}^{N}$, where Berger and Doggett [5] found $N = 25, 12, 9$ and 6 to be sufficient for shields of water, iron, tin and lead respectively. Thus the complete procedure is to generate histories (8.3.1) by sampling from the physical distributions governing the transition of energy and direction at each scattering, to compute for each such history the constants A_{ij} and thence the functions $P_i(t)$ from (8.3.9) to (8.3.12), and to use the average value of

$$\sum_{i=0}^{N} P_i(t) \quad (8.3.14)$$

as an approximate unbiased estimate of the penetration probability (8.3.13).

There are two advantages in this method. First, we do not introduce sampling variations due to sampling z as well as E and θ; thus, we improve the precision of the final result. This is line with the general maxim in § 1.1 that exact analysis should replace Monte Carlo sampling wherever possible. Berger and Doggett [5] found that the variance of the final result was reduced by a factor of about 10^4 at the cost of about 4 times as much computational labour. This is an impressive efficiency gain of about 2500. The second advantage of the method is that we may insert different values of $\zeta = t$ in (8.3.9) and thereby, from a single set of histories, estimate the probabilities of penetrating shields of various thicknesses t.

8.4 Use of control variates

It so happens that the exact expectation $\pi(t)$ of $\sum_{i=0}^{\infty} P_i(t, \infty, h)$, the density of photons crossing the plane $z = t$ in a semi-infinite medium, is known. There is a high correlation between $\sum_{i=0}^{\infty} P_i(t, t, h)$ and $\sum_{i=0}^{\infty} P_i(t, \infty, h)$ since the only difference arises from the particles that are scattered back across $z = t$ after having crossed it. It is therefore advantageous to use the latter as a control variate, observing the quantity

$$\sum_{i=0}^{n} \{P_i(t, t, h) - P_i(t, \infty, h)\} + \pi(t). \qquad (8.4.1)$$

This idea also is due to Berger and Doggett [5]. They found that this device gave a further reduction in variance by a factor of about 130 at the expense of about 60% extra computation. Thus the overall efficiency factor from the semi-analytic method combined with control variates is about 200,000.

8.5 Criticality problems in neutronics

The second problem of § 8.1 is rather different in character. For almost all assemblies of fissile and absorbent material, there is a constant μ and a corresponding density function $f(E, \boldsymbol{\omega}, \mathbf{r})$, such that a distribution of neutrons with density $A f(E, \boldsymbol{\omega}, \mathbf{r})$ at time zero will lead to a distribution with density $A f(E,\omega,\mathbf{r})e^{\mu t}$ at time t. (This is provided that t is not so large that the number of neutrons either grows so large that the supply of fissile material is exhausted or falls so low that to talk of a density becomes meaningless.)

According as μ is negative, zero, or positive, the system is *sub-critical, critical,* or *super-critical.* A complete solution of the problem consists in finding f and μ, but often one is content with μ, or even the sign of μ alone.

The standard method of solution by Monte Carlo methods is to start with some arbitrary distribution of neutrons and to track them for as long a time as possible. As time goes on, the distribution will tend towards the limiting form $A f(E,\omega,\mathbf{r})$, for some A, and when the distribution has settled down, the value of μ may be determined from the multiplication rate.

We must use weights, as described in § 8.2, to keep the number of tracks from increasing or decreasing too far. A simple estimator for μ is thus

$$\frac{1}{T_2 - T_1} \log\left(\frac{\text{total weight of paths at time } T_2}{\text{total weight of paths at time } T_1}\right), \qquad (8.5.1)$$

where T_1 and T_2 are large enough for the distribution to have settled down, and far enough apart for the change in weight to be significant. The easiest way to find out whether these rather vague conditions are satisfied is by trial and error, in other words by seeing whether different values of T_1 and T_2 lead to values of μ that are not significantly different.

We can improve on this by observing, instead of an actual change of weight, the instantaneous expected rate of change of weight. Let a particle, having weight w at time t, be travelling with velocity v (depending on E), let the local collision cross-section be σ_c and the absorption probability be α, as before, and let the fission probability be β and the expected number of particles to emerge from a fission be γ. Then the expected total weight of particles arising from this one particle up to the time $(t + \delta t)$, where δt is small, is

$$w + \sigma_c v \,\delta t\{-\alpha w + \beta(\gamma - 1)\,w\}, \qquad (8.5.2)$$

so that

$$\mathscr{E} \frac{d}{dt} \text{(total weight at time } t\text{)} = \sum_{\text{particles}} w\sigma_c v\{\beta(\gamma - 1) - \alpha\}. \tag{8.5.3}$$

It follows that we may estimate μ by the weighted mean of

$$\sigma_c v\{\beta(\gamma - 1) - \alpha\}, \tag{8.5.4}$$

over all the particles of the sample, observed at one sufficiently large time T. An even better estimator is got by averaging this weighted mean over several different values of T (or over a range of values, where this can be done analytically).

The time taken for this process to converge depends on how far the initial distribution of neutrons differs from f. This suggests another device for improvement; we may multiply the weights of the original neutrons and their respective descendants by appropriate factors to bring the initial distribution close to the final one; this will alter the weights appearing in the weighted mean, and should yield a better estimate of μ from the same observations. Essentially the same device was proposed for matrix problems in § 7.5.

8.6 The matrix method

If we do not ask for the value of μ, but merely whether it is positive or negative, we have still further scope for time-saving. The first thing we may do is to replace time as our independent variable, by 'generation number' n, the number of collisions appearing in the direct line of descent from an initial particle to one that is subsequently observed. If $g(E, \omega, \mathbf{r})$ is the conditional density function of those neutrons with density $f(E, \omega, \mathbf{r})$ which are just about to have a collision, it may be shown that a density of $Ag(E, \omega, \mathbf{r})$ at generation $n = 0$ will lead to a density of $Ag(E, \omega, \mathbf{r})\lambda^n$ at the nth generation, where λ is greater than, equal to, or less than 1 according as μ is greater than, equal to or less than 0.

A situation may arise in which we can divide the neutrons into classes on the basis of their energies, directions, or positions, so that the distribution of neutrons within each class settles down in very few generations, but their distribution between classes fluctuates for a much longer period. This is similar to the kind of problem in simple integration that is best solved by stratified sampling; here an appropriate technique is the *matrix method* developed by Morton and Kaplan and described in [6]. We give a brief outline of this method.

Divide all of $(E, \omega, \mathbf{r})$-space in a suitable manner into regions S_i. Then, every time a collision occurs, assign it to the region containing the point which describes the state of neutron just before that collision. Ignoring events which occur before the distributions in the regions have settled down, define N_{ij} to be the number of collisions in S_i of neutrons which themselves result from collisions in S_j, and define $C_{ij} = N_{ij} / \sum_i N_{ij}$. (If we are using weights we naturally redefine N_{ij} to be the sum of the appropriate weights.) Then C_{ij} is an estimator of K_{ij}, the expected number of collisions in S_i arising from one collision in S_j, and the dominant latent root of the matrix $\{K_{ij}\}$ is λ. We may thus estimate λ by finding the dominant latent root of $\{C_{ij}\}$ as soon as its elements have settled down, as they do much earlier than do the actual numbers N_{ij}.

8.7 Semi-analytic and antithetic techniques in criticality calculations

In (8.5.3) we improved our estimator by considering the expected rate of increase of neutrons per unit time. The same device can be employed when considering the situation generation by generation: to estimate λ we then require the expected number of neutrons in the next generation per neutron in the present generation.

Suppose that we have reached a situation in which we are sampling neutrons from equilibrium distribution $f(E, \omega, \mathbf{r})$: that is to say, f is the dominant eigenvector. The straightforward approach would be to sample a neutron from this distribution, thus according it values of $E, \omega, \mathbf{r}$, and then to determine from the exponential distribution a number s (such that the next collision it suffers is at $\mathbf{r} + s\omega$) and from the fission distribution a number $c = c(s)$ representing the number of secondary neutrons arising from this collision. The multiplication rate λ would then be the expected value of $c(s)$, and we could use the observed value of $c(s)$ to estimate λ. However, it may be possible to calculate this expectation analytically. The expectation is

$$\int c(s)\, e^{-s}\, ds, \qquad (8.7.1)$$

if the mean free path is unity; and it is then a question of being able to work out the integral (8.7.1). If the mean free path depends upon the region of the medium, we shall need a more complicated factor in place of e^{-s} in the integral; but, depending upon the context, we may still be able to perform the integration. We may even be able to do better than this, and take an expectation not only over values of s but also over values of ω say. The general principle is always to do as much of the work as we can analytically, for this will sharpen the precision of the estimators.

Rather than talk in terms of generalities, let us consider a specific example [7], [8] in which the physical situation is greatly simplified for the sake of clarity of exposition.

Suppose that we have a sphere of homogeneous fissile material *in vacuo*. Choose the units of length so that the mean free path within the sphere is unity. We shall suppose that the only event which takes place upon collision is fission, and that exactly 2 neutrons emerge isotropically from each such fission. Thus $c(s) = 2$ within the sphere and $c(s) = 0$ outside the sphere; and the integral (8.7.1) becomes $2(1 - e^{-t})$, where t is the distance from r along ω to the surface of the sphere. In the practical case considered in [7], the radius of the sphere was 1·10. To arrive at the equilibrium distribution g, 25 neutrons were started from the centre of the sphere, and they and their descendant neutrons were tracked for 19 generations. After about 11 generations, the distribution seemed to have settled down; and here we shall consider some of the neutrons in the 17th generation. The 17th generation actually consisted of 83 neutrons, from which we draw 10. Table 8.1 shows in its first column, the distances r from the centre of the sphere of these 10 neutrons. Since the secondaries emerge isotropically, t is simply the distance from the point r to the boundary of the sphere, measured along a randomly selected direction. Sampling 10 such directions we arrive at the values of $2(1 - e^{-t})$ in the second column; and our estimate of λ is simply the mean (0·985) of these.

Table 8.1

r	$2(1-e^{-t})$	$2-e^{-t}-e^{-t'}$	$K(r)$
0·66	1·466	1·178	1·176
0·53	0·986	1·244	1·238
0·19	1·354	1·316	1·322
0·47	0·956	1·260	1·260
1·08	0·698	0·462	0·660
0·92	0·892	0·982	0·966
0·89	0·902	0·954	1·000
0·69	1·128	1·152	1·160
0·90	1·278	0·968	0·988
0·98	0·190	0·820	0·884
Mean	0·985 ± 0·116	1·034 ± 0·081	1·065 ± 0·064

The entry $\pm 0·116$ is the standard error of this mean calculated from the variance of the 10 values of $2(1-e^{-t})$.

Evidently there is a good deal of scatter in these 10 values of $2(1-e^{-t})$, and we cannot expect 0·985 to be a very good estimate of λ. We can do better by using antithetic variates (§ 5.6). Each t is the distance to the boundary in the random direction. We choose t' to be distance in the diametrically opposite direction. The t and t' are negatively correlated, and a simple antithetic estimator is

$$\tfrac{1}{2}[2(1-e^{-t})+2(1-e^{-t'})] = 2-e^{-t}-e^{-t'}, \qquad (8.7.2)$$

which appears in the third column. Clearly the scatter amongst these 10 values is a good deal less, and we expect their mean and standard error (1·034 ± 0·081) to be a better estimate of λ.

We can go further by integrating analytically over all directions ω of the secondary neutrons. Let θ denote the angle between direction ω and the radius vector to the point of collision. Since we are assuming isotropy, $\cos\theta$ is uniformly distributed between -1 and $+1$, and so has the frequency element $\tfrac{1}{2}d(\cos\theta)$. Simple trigonometry shows that

$$t = -r\cos\theta + \sqrt{(a^2-r^2\sin^2\theta)}, \qquad (8.7.3)$$

where $a = 1 \cdot 10$ is the radius of the sphere. The expected value of $2(1 - e^{-t})$, averaged over ω, is accordingly

$$K(r) = \int_{\theta=0}^{\pi} 2[1 - \exp\{r\cos\theta - \sqrt{(a^2 - r^2\sin^2\theta)}\}]\tfrac{1}{2}d(\cos\theta)$$

$$= 2 - \frac{1}{2r}\{(1 + a + r)e^{-a+r} - (1 + a - r)e^{-a-r} -$$

$$- (a^2 - r^2)[Ei(-a-r) - Ei(-a+r)]\}, \qquad (8.7.4)$$

where

$$Ei(-x) = \int_{x}^{\infty} e^{-u} du/u \qquad (8.7.5)$$

is the exponential integral. The values of $K(r)$ corresponding to the 10 sample values of r appear in the last column of the above table and their average gives $1 \cdot 065$ as an estimate of λ.

The problem under consideration is, of course, a very simple neutron diffusion problem, so simple indeed that it can be solved theoretically by the so-called extrapolated end-point method [9]. The exact value of λ is $1 \cdot 0654$. It is an exceptional fluke that the third column of the table should give so correct a result on a sample of only 10 neutrons. Serious calculations would use a larger sample.

We may compare the foregoing estimates of λ with one obtained by crude Monte Carlo tracking. In this tracking, 12 of the 20 neutrons released by these 10 collisions had further collisions within the sphere, the remaining 8 escaping. The estimate of λ is accordingly $12/10 = 1 \cdot 2$. The probability of remaining in the sphere for a further collision is $12/20 = 0 \cdot 6$; and the standard deviation of the number remaining is given by the binomial formula $\sqrt{(12 \times 0 \cdot 6 \times 0 \cdot 4)} = 1 \cdot 69$. The standard error of the estimate of λ will be one tenth of this. Thus crude tracking gives $\lambda = 1 \cdot 2000 \pm 0 \cdot 169$.

There is, of course, no need to confine the estimation to neutrons in a single generation. Suppose that there are n neutrons in the sphere at the ith generation. Suppose also that the system has settled down to its equilibrium distribution, and write p for the probability that a neutron, chosen at random from this distribution, will have its next collision within the sphere. Then $\lambda = 2p$, since this neutron will produce 2 secondaries. If there are n_i neutrons in the sphere immediately before the collisions of the generation, there will be $2n_i$ immediately after. The likelihood of n_{i+1} of these remaining in the sphere for a collision in the $(i+1)$th generation will be

$$L_i = \binom{2n_i}{n_{i+1}} p^{n_{i+1}}(1-p)^{2n_i-n_{i+1}} \qquad (8.7.6)$$

according to the binomial distribution. Hence, given the number n_j in the jth generation, the conditional likelihood of $n_{j+1}, \ldots, n_k$ in the $(j+1)$th, $\ldots$, kth generations is

$$L = \prod_{i=j}^{k-1} \binom{2n_i}{n_{i+1}} p^{n_{i+1}}(1-p)^{2n_i-n_{i+1}}. \qquad (8.7.7)$$

The maximum likelihood estimator (§ 2.4) of p is the root of the equation

$$\partial \log L/\partial p = 0, \qquad (8.7.8)$$

namely,

$$\hat{p} = (N-n_j)/2(N-n_k), \qquad (8.7.9)$$

where

$$N = n_j + n_{j+1} + \ldots + n_k. \qquad (8.7.10)$$

The variance of this estimator is $-\partial^2 \log L/\partial p^2$ evaluated at $p = \hat{p}$, and this gives

$$\mathrm{var}\,\hat{p} = (N-n_j)(N+n_j-2n_k)/8(N-n_k)^3. \qquad (8.7.11)$$

The maximum likelihood estimator of λ and its standard error are accordingly

$$\hat{\lambda} = \frac{(N-n_j)}{(N-n_k)} \pm \sqrt{\left[\frac{(N-n_j)(N+n_j-2n_k)}{2(N-n_k)^3}\right]}. \qquad (8.7.12)$$

It is informative to put this estimator in another form. We have identically

$$\hat{\lambda} = \frac{(N-n_j)}{(N-n_k)} = \left(\sum_{i=j}^{k-1} n_j \frac{n_{i+1}}{n_i}\right)\bigg/\left(\sum_{i=j}^{k-1} n_i\right); \qquad (8.7.13)$$

that is to say λ is the weighted mean of the individual estimators n_{i+1}/n_i of the multiplication rate generation by generation, the weights being n_i, the number of neutrons in the parent generation, and therefore inversely proportional to the variances of these individual estimates. A similar weighting procedure is appropriate to the case when we are combining estimates from different generations in a more sophisticated analysis, such as the semi-analytic or antithetic method.

In the Monte Carlo tracking from the 11th to the 19th generations, Hammersley and Morton [8] found $n_{11} = 51$, $n_{19} = 94$, $N = 665$. By (8.7.12) this gives

$$\hat{\lambda} = 1 \cdot 076 \pm 0 \cdot 030. \tag{8.7.14}$$

CHAPTER 9

Problems in Statistical Mechanics

9.1 Markov chains

Markov chains have appeared several times in Chapters 7 and 8 of this book, and it is now time that we gave them formal recognition.

We shall concern ourselves only with chains whose transitions occur at discrete times. Let these times be labelled consecutively 1, 2, 3, ..., let the system have a finite or countable set of possible states S_1, S_2, ..., and let X_t be the state that it is in at time t. X_t is then a random variable, and we may define the conditional probabilities

$$P(X_t = S_j | X_{t_1} = S_{i_1}, X_{t_2} = S_{i_2}, \ldots, X_{tn} = S_{i_n}). \quad (9.1.1)$$

The system is a *Markov chain* if the distribution of X_t is independent of all previous states except for its immediate predecessor X_{t-1}, or, more formally,

$$P(X_t = S_j | X_{t-1} = S_{i_{t-1}}, \ldots, X_2 = S_{i_2}, X_1 = S_{i_1})$$
$$= P(X_t = S_j | X_{t-1} = S_{i_{t-1}}). \quad (9.1.2)$$

In the cases that we consider, the probability (9.1.2) is independent of t, and we have a chain with stationary transition probabilities, defined by

$$p_{ij} = p(S_i \to S_j) = P(X_t = S_j | X_{t-1} = S_i). \quad (9.1.3)$$

By the use of density functions in place of probabilities, this may all be extended to systems in which the possible states form a continuum.

For example, the random walk defined in § 7.4 is a Markov chain, whose states comprise all possible positions of the particle within the boundary, together with a special 'absorbing' state that the system enters as soon as the particle hits the boundary and remains in for ever after. On the other hand the 'self-avoiding' walks to be discussed in Chapter 10 are not Markov chains when described in the same manner, since the condition that the system never occupies the same state twice makes the distribution of the current state depend on the whole past history of the walk.

If f is any function defined on the set of states $\{S_i\}$, this defines a random variable

$$y_t = f(X_t), \tag{9.1.4}$$

and we shall in our applications be looking at variables of the form

$$Y_t = \frac{1}{t} \sum_{s=1}^{t} y_s, \tag{9.1.5}$$

which is the average value of f over the first t states of a realization of the Markov chain. A special case of (9.1.5) is the variable

$$\left. \begin{array}{l} \Delta_t(i) = \frac{1}{t} \sum_{s=1}^{t} \delta_s(i) \\[2mm] \text{where} \qquad \delta_t(i) = 1 \text{ if } X_t = S_i, = 0 \text{ if } X_t \neq S_i \end{array} \right\} \tag{9.1.6}$$

measuring the frequency with which the state S_i occurs; we may, if we prefer, use as the definition of Y_t,

$$Y_t = \sum_i \Delta_t(i) f(S_i). \tag{9.1.7}$$

If we write

$$p_{ij}^{(n)} = P(X_t = S_j | X_{t-n} = S_i), \tag{9.1.8}$$

defining the n-step transition probabilities, we have

$$p_{ij}^{(1)} = p_{ij}, \ p_{ij}^{(n+1)} = \sum_k p_{ik}^{(n)} p_{kj}, \tag{9.1.9}$$

or, writing $\mathbf{P}$ and $\mathbf{P}^{(n)}$ for the matrices with elements p_{ij}, $p_{ij}^{(n)}$ respectively,

$$\mathbf{P}^{(n)} = \mathbf{P}^n. \tag{9.1.10}$$

The matrix $\mathbf{P}$ is *stochastic*†, that is to say

$$p_{ij} \geqslant 0 \quad \text{and} \quad \sum_j p_{ij} = 1, \tag{9.1.11}$$

and it is easily shown that the same is true of the matrix $\mathbf{P}^n$.

We may also define the *first-passage* probabilities,

$$f_{ij}^{(n)} = P(X_t = S_j, X_{t-1} \neq S_j, \ldots, X_{t-n+1} \neq S_j | X_{t-n} = S_i), \tag{9.1.12}$$

and show that they satisfy

$$f_{ij}^{(1)} = p_{ij} f_{ij}^{(n+1)} = p_{ij}^{(n+1)} - \sum_{r=1}^{n} f_{ij}^{(r)} p_{jj}^{(n-r+1)}, \tag{9.1.13}$$

The *mean first passage times* (*mean recurrence times* if $i = j$) are defined by

$$m_{ij} = \sum_{n=1}^{\infty} n f_{ij}^{(n)}, \tag{9.1.14}$$

provided that

$$\sum_{n=1}^{\infty} f_{ij}^{(n)} = 1. \tag{9.1.15}$$

If $\sum_{n=1}^{\infty} f_{ii}^{(n)} = 1$, the state S_i is called *recurrent*‡; if then $m_{ii} < \infty$ it is called *positive*, if $m_{ii} = \infty$ the state is *null*. If $p_{ii}^{(n)} \neq 0$ only when n is a multiple of d, the state S_i has *period* d; if $d = 1$, the state is *aperiodic*.

If S_i and S_j are mutually accessible, that is, if there exist numbers m and n such that

$$p_{ij}^{(m)} \neq 0, \quad p_{ji}^{(n)} \neq 0, \tag{9.1.16}$$

then S_i and S_j belong to the same *class*; it may be proved that the

---

† If $\sum\limits_{j} p_{ij} < 1$, introduce an additional state S_ω, with $p_{i\omega} = 1 - \sum\limits_{j} p_{ij}$, $p_{\omega j} = 0$, $p_{\omega\omega} = 1$; the extended matrix is then stochastic.

‡ A great variety of terminology exists in the literature, and authors (e.g. Feller [1]) even permute the meanings of various adjectives from one edition to another. As a result, all such words as 'recurrent', 'null', 'ergodic', 'transient', 'persistent', etc., are so obscured by hopeless confusion that nobody knows what they mean in any given context. Really, somebody ought to start afresh with a completely new set of adjectives, but we do not have the audacity to do so.

states of a class are either all non-recurrent, or all positive, or all null, and that all have the same period. We shall be concerned only with *irreducible* chains, all of whose states are in the same class, and primarily with chains which are finite (in their numbers of states) and aperiodic. In the finite case it can be shown that all the states must be positive.

The following statements are true of irreducible chains. For proofs and more general statements we refer the reader to the books by Feller [1] chapter 15, and Chung [2] part I §§ 15–16.

I.   If the states are positive and aperiodic, then

$$\lim_{n \to \infty} p_{ij}^{(n)} = \pi_j = m_{jj}^{-1} \qquad (9.1.17)$$

and $\{\pi_j\}$ is the unique set of numbers satisfying

$$\left. \begin{aligned} \pi_j > 0, \ \sum_j \pi_j = 1, \\ \pi_j = \sum_i \pi_i p_{ij}. \end{aligned} \right\} \qquad (9.1.18)$$

II.   If numbers $\{\pi_j\}$ exist satisfying (9.1.18), and if states are aperiodic, then these numbers are unique, the states are positive, and (9.1.17) holds.

III.   If the states are positive, or if (9.1.18) is satisfied, then $\Delta_t(j)$ tends to m_{jj}^{-1} with probability one as $t \to \infty$.

IV.   If the conditions for III hold, and if

$$m_{ii}^{(2)} = \sum_{n=1}^{\infty} n^2 f_{ii}^{(n)} < \infty, \qquad (9.1.19)$$

then the expectation of $(\Delta_t(j) - m_{jj}^{-1})^2$ is $O(t^{-1})$.

V.   If the conditions for IV hold, and if the expectation of the square of the sum of the values of $|y_t|$ between consecutive recurrences of S_i is finite, or in particular, if f is bounded, then the expectation of

$$\{Y_t - \sum_j m_{jj}^{-1} f(S_j)\}^2 \text{ is } O(t^{-1}).$$

In the case of a finite irreducible chain, all the conditions for I–V are automatically satisfied, except that the states may be periodic. Notice that if the chain is reducible the solution of (9.1.18) is not unique.

9.2 Problems in equilibrium statistical mechanics

Statistical mechanics [3] [4] studies physical systems consisting of a large number of identical components, such as the molecules in a gas. Its use is justified on one of two grounds: either one allows chance to play a fundamental part in the laws of physics, or one says that the laws are deterministic but it is impossible to collect enough facts to be able to apply them deterministically, so that one turns to probability to cover up one's ignorance.

If a classical system is in *thermal equilibrium* with its surroundings, and is in a state S with energy $E(S)$, then the probability density in phase-space of the point representing S is proportional to

$$e^{-\beta E(S)}, \tag{9.2.1}$$

where $\beta = (kT)^{-1}$, T is the absolute temperature of the surroundings, and k is Boltzmann's constant. According to the ergodic theory, the proportion of time that the system spends in state S is also proportional to (9.2.1). If the system is observed at a random time, the expectation† $\langle f \rangle$ of any state-function $f(S)$ is thus

$$\langle f \rangle = \frac{\int f(S)\, e^{-\beta E(S)}\, dS}{\int e^{-\beta E(S)}\, dS}, \tag{9.2.2}$$

with some restrictions on the form of f. The Monte Carlo method tries to evaluate (9.2.2) for systems for which this cannot be done analytically.

9.3 Metropolis' method for evaluating $\langle f \rangle$

One could in theory evaluate (9.2.2) by crude Monte Carlo estimation of the two integrals, possibly using the same random numbers in

---

† In statistical mechanics, it is conventional to write $\langle f \rangle$ for the expectation of a quantity f. We adopt this convention here, but restrict its use to expectations over the Boltzmann distribution (9.2.1).

each case. However, this method breaks down in practice because the exponential factor means that the significant part of the integral is concentrated in a very small region of phase-space. We therefore want to turn to importance sampling, and to generate states with a probability density of

$$\pi(S) = e^{-\beta E(S)} \Big/ \int e^{-\beta E(S)} dS. \qquad (9.3.1)$$

Then f is an unbiased estimator of the whole expression (9.2.2).

On the face of it this is impossible, for the same reason as in § 5.4, namely that we do not know the denominator of (9.3.1). We may get round this, however, by a device first announced by Metropolis and his collaborators [5] in 1953.

Suppose, for simplicity, that the phase-space is discrete, so that the integrals in (9.2.2), (9.3.1) are in fact summations over discrete states S_j. If we can find an irreducible aperiodic Markov chain with transition probabilities p_{ij} such that

$$\pi_j = \pi(S_j) = \sum_i \pi_i p_{ij}, \qquad (9.3.2)$$

then (9.1.18) is satisfied and (by II of § 9.1) the Markov chain has the unique limit distribution π_j; we may then hope that f is such that (by V) the mean value Y_t of f over t consecutive states in a realization of the Markov chain tends to $\langle f \rangle$ with an error that is $O(t^{-1/2})$ as t increases. The point to notice about (9.3.2) is that it involves only the relative values of $\pi(S)$, i.e. the ratios π_i/π_j, so that one never needs to evaluate the denominator of (9.3.1).

Metropolis et $al.$ contrived to satisfy (9.3.2) as follows. Consider an arbitrary symmetric Markov chain, that is to say one with a symmetric matrix $\mathbf{P}^*$ of transition probabilities. The elements of this matrix satisfy

$$p_{ij}^* \geqslant 0, \quad \sum_j p_{ij}^* = 1, \quad p_{ij}^* = p_{ji}^*. \qquad (9.3.3)$$

The first two conditions are those satisfied by any stochastic matrix, as in (9.1.11), while the third is the condition of symmetry. We now define a set of quantities p_{ij} in terms of p_{ij}^* and the $known$ ratios π_i/π_j;

and we shall prove that these p_{ij} are the elements of a stochastic matrix satisfying (9.3.2). If $i \neq j$ we define

$$p_{ij} = \begin{cases} p_{ij}^* \pi_j / \pi_i & \text{if } \pi_j / \pi_i < 1 \\ p_{ij}^* & \text{if } \pi_j / \pi_i \geqslant 1. \end{cases} \qquad (9.3.4)$$

If $i = j$ we define

$$p_{ii} = p_{ii}^* + \sum_j{}' p_{ij}^* (1 - \pi_j / \pi_i), \qquad (9.3.5)$$

where $\sum'$ is taken over all values of j such that $\pi_j / \pi_i < 1$. In the first place, since each $\pi_i \geqslant 0$, all $p_{ij} \geqslant 0$. In the second place, writing $\sum_j{}''$ for summation over all values of $j \neq i$ such that $\pi_j / \pi_i \geqslant 1$, we have

$$\sum_j p_{ij} = p_{ii}^* + \sum_j{}' p_{ij}^* (1 - \pi_j / \pi_i) + \sum_j{}' p_{ij}^* \pi_j / \pi_i + \sum{}'' p_{ij}^*$$

$$= p_{ii}^* + \sum_j{}' p_{ij}^* + \sum_j{}'' p_{ij}^* = p_{ii}^* + \sum_{j \neq i} p_{ij}^*$$

$$= \sum_j p_{ij}^* = 1. \qquad (9.3.6)$$

Thus the p_{ij} satisfy (9.1.11) and are therefore elements of a stochastic matrix.

Next if i and j are two suffices for which $\pi_i = \pi_j$ we have by (9.3.3) and (9.3.4), the latter used once as it stands and once with i and j reversed,

$$p_{ij} = p_{ij}^* = p_{ji}^* = p_{ji}, \qquad (9.3.7)$$

and therefore, since $\pi_i = \pi_j$,

$$\pi_i p_{ij} = \pi_j p_{ji}. \qquad (9.3.8)$$

On the other hand if $\pi_j < \pi_i$, we have again by (9.3.3) and (9.3.4) used twice,

$$p_{ij} = p_{ij}^* \pi_j / \pi_i = p_{ji}^* \pi_j / \pi_i = p_{ji} \pi_j / \pi_i; \qquad (9.3.9)$$

and this again gives (9.3.8). Similarly (9.3.8) results from the supposition that $\pi_i < \pi_j$. Consequently (9.3.8) holds for all values of i, j. Finally by (9.3.8) and (9.3.6)

$$\sum_i \pi_i p_{ij} = \sum_i \pi_j p_{ji} = \pi_j \sum_i p_{ji} = \pi_j; \qquad (9.3.10)$$

and this completes the proof that the p_{ij} satisfy (9.3.2).

Next let us see how to apply this result. We take **P*** to be the transition matrix of a Markov chain on the states $\{S_i\}$. Suppose that we are currently in state S_i. We use $\mathbf{P}^*$ to pick a state S_j; the transition probability of this is p_{ij}^*. Having chosen S_j in this way, we calculate the ratio π_j/π_i. If $\pi_j/\pi_i \geqslant 1$, we accept S_j as the new state. On the other hand if $\pi_j/\pi_i < 1$, then with probability π_j/π_i we accept S_j as the new state, and with the remaining probability $1 - \pi_j/\pi_i$, we take the old state S_i to be the new state. This procedure gives the transition probabilities p_{ij} defined in (9.3.4) and (9.3.5).

If $\pi_j > 0$ for all j, $\mathbf{P} = (p_{ij})$ will represent an irreducible aperiodic Markov chain whenever $\mathbf{P}^* = (p_{ij}^*)$ does, because there will be a positive probability of following under $\mathbf{P}$ any given finite sequence of states that can arise with positive probability under $\mathbf{P}^*$. Should there be states S_j for which $\pi_j = 0$, the above does not necessarily hold, and an irreducible $\mathbf{P}^*$ may lead to a reducible $\mathbf{P}$; however, we have plenty of latitude in our choice of $\mathbf{P}^*$, and shall usually be able to prevent this happening.

In the case of the Boltzmann distribution (9.3.1), we have

$$\pi_j/\pi_i = e^{-\beta E(S_j) + \beta E(S_i)} = e^{-\beta \Delta E}, \tag{9.3.11}$$

where ΔE is the difference of the energies in the two states. To summarize, our sampling procedure is the following.

If $X_t = S_i$ let X_t^* be a state selected from a distribution such that

$$P(X_t^* = S_j | X_t = S_i) = P(X_t^* = S_i | X_t = S_j). \tag{9.3.12}$$

Let $\Delta E = E(X_t^*) - E(X_t)$. Then if $\Delta E \leqslant 0$ we take $X_{t+1} = X_t^*$, while if $\Delta E > 0$ we take

$$X_{t+1} = \begin{cases} X_t^* \text{ with probability } e^{-\beta \Delta E} \\ X_t \text{ with probability } 1 - e^{-\beta \Delta E}. \end{cases}$$

The transition probabilities p_{ij} thus defined satisfy not only (9.3.2) but the stronger condition (9.3.8). The chain so defined is aperiodic and the only doubts are whether the conditions for IV and V of § 9.1 are satisfied, and whether the chain is irreducible.

Assuming that these doubts have been allayed, there remains a practical problem arising from the fact that we can realize only a finite number of steps of the chain. Wood and Parker [6] call a chain *quasi-ergodic* if there are sets of states of low energy (and therefore of high probability) such that the chance of passing from one set to another in the course of the realization is small. Then it is quite likely that Y_t may take values at some distance from $\langle f \rangle$, while giving every sign of having converged. It is a wise precaution to run two or more simultaneous independent realizations of the Markov chain with completely different initial states. We may then use the difference between the respective values of Y_t for a convergence test. In any case, there is a good deal of room for further research into how best to choose the arbitrary symmetric P^* to achieve rapid convergence and small variances for the final estimates.

Formally, it is easy to extend this process to continuous phase-space, when (9.3.2) for instance, goes into

$$\pi(S) = \int \pi(S')p(S' \to S)\,dS' \tag{9.3.13}$$

where p is now a conditional density.

9.4 Equations of state

Metropolis originally proposed his method in connexion with investigations of the behaviour of a liquid or a dense gas, represented by interacting molecules confined to a box. See [5], also [7], [8], [9] and for a review [10].

One has N molecules centred at points $\mathbf{x}_1, \mathbf{x}_2, \ldots, \mathbf{x}_N$, with an energy of interaction

$$E(S) = \sum_i \sum_j V(|\mathbf{x}_i - \mathbf{x}_j|), \tag{9.4.1}$$

and wants to estimate such parameters as the mean square separation of a molecule from its nearest neighbour. The commonest forms of interaction considered are the hard-sphere potential

$$V(r) = \begin{cases} 0, & r > \sigma \\ \infty, & r < \sigma \end{cases} \tag{9.4.2}$$

and the Lennard-Jones potential

$$V(r) = 4\epsilon\{(\sigma/r)^{12} - (\sigma/r)^6\}. \tag{9.4.3}$$

The method is as outlined in § 9.3, X_t being generated from X_t^* by selecting a molecule at random and subjecting it to a translation $\mathbf{v}$ chosen uniformly at random from a cube $-\alpha \leqslant \mathbf{v} \leqslant \alpha$.

This is a case of a continuous phase-space, for which the validity of the general process has not been established. In this particular instance, it is not difficult to carry out the analysis, except that, when the molecules are very densely packed and the hard-sphere potential is used, it may not be evident whether the chain is irreducible; if the cube-size α is taken as small, the Monte Carlo process resembles a possible physical process, in which the molecules may conceivably not have room to squeeze past one another. This difficulty never arises when α is large, but practical considerations sometimes favour a small α.

The process is not feasible for more than a few hundred molecules, but a good approximation to a larger system is got by the use of *periodic boundary conditions*, under which all co-ordinates are interpreted modulo 1 so that each molecule stands for an infinite lattice of molecules. To avoid the excessive labour of computing (9.4.3) for every pair of molecules in the system, Wood and Parker [6] use an analytic approximation for the effect of all distant pairs, using the exact formula for small values of r only.

We should mention the work of Alder and Wainwright [11]. Theirs is really a simulation method, since they take an initial configuration of molecules with (usually) equal energies and random directions of motion, and follow their subsequent motion by the ordinary methods of dynamics from collision to collision, for several thousand collisions.

9.5 Order-disorder phenomena

Another successful application of Metropolis' method has been to the Ising model [12] variously representing the behaviour of a substitutional alloy [13], a ferromagnet [14], [15], [16], or other co-operative phenomena. The model consists of a lattice of sites, each occupied by one or other type of atom (in case of an alloy) or by a spin in one of two orientations (in the case of a magnet). The energy then depends upon the number of sites and the number of pairs of adjacent sites occupied in the various possible manners (e.g. a pair of sites could be occupied by AA, AB, BA or BB, where A and B represent the two components of an alloy).

Now we have a genuinely discrete phase-space, so that the theory is rigorous. The number of sites that we can handle, however, is again restricted, and we need periodic boundary conditions again as in §9.4.

In the case of an alloy, we probably want to keep the proportions of atoms of each type fixed, so that we may generate X_t^* from X by exchanging two atoms on a pair of adjacent sites chosen at random, or perhaps simply on any two random sites. In the case of a magnet, the number of spins in each orientation contributes to the energy, so that it seems preferable to choose one site at random and to reverse the spin on it to obtain X_t^*. It has been suggested that one would do better to choose the sites for exchange or reversal systematically, rather than at random. This calls for a slight modification of the theory. If P_1, P_2, ..., P_R are the transition-probability matrices appropriate to the R possible choices, in the order in which they are systematically chosen, and if

$$P = P_1 P_2 \ldots P_R, \qquad (9.5.1)$$

then one justifies the procedure by showing that (9.3.8) is satisfied by each P_r, and therefore by P, and then showing that P is irreducible, although the individual P_r are not. The results therefore hold good, provided that we regard R steps of the process as constituting a single step of the Markov chain.

The most interesting feature of the Ising model is the existence of a transition temperature or *Curie point* T_c, at which the properties of an infinite lattice show a radical change; at all temperatures below T_c there is a condition of *long-range order*, under which the configurations most likely to occur display a degree of regularity that is entirely absent at higher temperatures. The finite lattices that the Monte Carlo method is compelled to use show traces of this phenomenon (see [14]), and so a rough value of T_c may be guessed at by inspection of the graph of some $\langle f \rangle$ as a function of T. We should like to have a method of estimating T_c directly from a single Monte Carlo experiment, but no suitable process has yet been discovered.

It should be remembered that Metropolis' Monte Carlo process is not intended to simulate the behaviour of the model in any respect other than its distribution of states; the sequence in which the states follow one another has no significance at all.

9.6 Quantum statistics

It may sometimes be worth while, or even essential, to transform the physical model before applying the process of § 9.3. In quantum statistical mechanics, for instance, it is essential. In place of (9.2.2) one has the expression

$$\langle \mathbf{F} \rangle = \frac{\text{trace } \{\mathbf{F} \, e^{-\beta \mathbf{H}}\}}{\text{trace } \{e^{-\beta \mathbf{H}}\}}, \tag{9.6.1}$$

where $\mathbf{F}$ and $\mathbf{H}$ are linear operators, $\mathbf{H}$ being the Hamiltonian of the system. Since one can no longer deal with the energy $E(S)$ of an individual state, Metropolis' method as it stands cannot be applied.

One way round this obstacle is to diagonalize the operators. Suppose that the states of the system are discrete so that $\mathbf{F}$ and $\mathbf{H}$ can be represented by (Hermitian) matrices. If we can find a transformation $\mathbf{T}$ to a new set of states S^* such that

$$|S_i\rangle = \sum_j T_{ij} |S_j^*\rangle, \tag{9.6.2}$$

and such that $\mathbf{F}^* = \overline{\mathbf{T}}'\mathbf{F}\mathbf{T}$ and $\mathbf{H}^* = \overline{\mathbf{T}}'\mathbf{H}\mathbf{T}$ are diagonal matrices, then we can use the previous method, taking

$$f(S_j^*) = F_{jj}^*, \quad E(S_j^*) = H_{jj}^*. \tag{9.6.3}$$

The usefulness of this approach depends on our ability to find and carry out the transformation $\mathbf{T}$.

An alternative way is that followed in [17] and [18]. Suppose that

$$\mathbf{H} = \mathbf{H}_0 + \sum_{i=1}^{N} \mathbf{H}_i, \tag{9.6.4}$$

where $\mathbf{H}_0$ commutes with every $\mathbf{H}_i$. Then, by expansion of the exponential in the usual series,

$$\begin{aligned}
\text{trace } \{\mathbf{F} \, e^{-\beta \mathbf{H}}\} &= \text{trace } \{\mathbf{F} \, e^{-\beta \mathbf{H}_0} e^{-\beta \sum_i \mathbf{H}_i}\} \\
&= \sum_{r=0}^{\infty} \frac{(-\beta)^r}{r!} \text{trace } \left\{ \mathbf{F} \, e^{-\beta \mathbf{H}_0} \left(\sum_i \mathbf{H}_i \right)^r \right\} \\
&= \sum_{r=0}^{\infty} \frac{(-\beta)^r}{r!} \sum_{\gamma_r} \text{trace } \left\{ F e^{-\beta \mathbf{H}_0} \mathbf{H}_{i_1} \mathbf{H}_{i_2} \dots \mathbf{H}_{i_r} \right\},
\end{aligned} \tag{9.6.5}$$

where γ_r runs over all possible sequences $i_1, i_2, \ldots, i_r$ of numbers in the range from 1 to N. Now the trace of an operator is a real number, so that we may write

$$\langle \mathbf{F} \rangle = \sum_r \sum_{\gamma_r} f(\gamma_r)\, \pi(\gamma_r), \qquad (9.6.6)$$

where the summation runs over all the sequences γ_r of all lengths from zero upwards, where

$$f(\gamma_r) = \frac{\text{trace}\,\{\mathbf{F}\,e^{-\beta \mathbf{H}_0}\mathbf{H}_{i_1}\mathbf{H}_{i_2}\ldots\mathbf{H}_{i_r}\}}{\text{trace}\,\{e^{-\beta \mathbf{H}_0}\mathbf{H}_{i_1}\mathbf{H}_{i_2}\ldots\mathbf{H}_{i_r}\}}, \qquad (9.6.7)$$

and where

$$\pi(\gamma_r) = A(-\beta)^r\,\text{trace}\,\{e^{-\beta \mathbf{H}_0}\mathbf{H}_{i_1}\mathbf{H}_{i_2}\ldots\mathbf{H}_{i_r}\}/r! \qquad (9.6.8)$$

(A being a constant such that $\sum_r \sum_{\gamma_r} \pi(\gamma_r) = 1$).

Provided that the traces are easily calculated, we can then evaluate (9.6.6) by Metropolis' method. Let $f_r (r = 0, 1, \ldots)$ be any sequence of probabilities and define

$$\lambda_r = \frac{1 - f_r}{r f_{r-1}}; \qquad (9.6.9)$$

let $p(i)$ ($i = 1, 2, \ldots, N$) be a set of probabilities with $\sum_i p(i) = 1$, and define

$$\tau(\gamma_r) = \frac{(-\beta)^r \lambda_1 \lambda_2 \ldots \lambda_r\,\text{trace}\,\{e^{-\beta \mathbf{H}_0}\mathbf{H}_{i_1}\mathbf{H}_{i_2}\ldots\mathbf{H}_{i_r}\}}{p(i_1)p(i_2)\ldots p(i_r)}. \qquad (9.6.10)$$

We construct a Markov chain whose states are the sequences γ_r. If $X_t = \gamma_r = i_1 i_2 \ldots i_r$, we first select the 'forward' direction with probability f_r or the 'backward' direction with probability $1 - f_r$. In the former case we take $X_t^* = \gamma_r i = i_1 i_2 \ldots i_r i$, where i is chosen from the distribution with frequencies $p(i)$. In the latter case, we take $X_t^* = i_1 i_2 \ldots i_{r-1}$, unless $r = 0$ when $X_t^* = X_t = \gamma_0$ (the empty sequence). Then if $\tau(X_t^*) \geq \tau(X_t)$ we take $X_{t+1} = X_t^*$, while if $\tau(X_t^*) < \tau(X_t)$ we take

$$X_{t+1} = \begin{cases} X_t^* & \text{with probability } \tau(X_t^*)/\tau(X_t) \\ X_t & \text{with probability } 1 - \tau(X_t^*)/\tau(X_t). \end{cases}$$

This is an exact analogue of Metropolis' method and leads to a Markov chain with the correct limit distribution (9.6.8); we may therefore estimate $\langle F \rangle$ by the average value of $f(\gamma_r)$ over the sequences in a realization of the chain.

This process suffers from the drawback that, while the end of the sequence γ_r is constantly changing, the earlier part tends to stay fixed, so that the chain is in danger of being only quasi-ergodic. A cure for this is to arrange that before every step the sequence is permuted from $i_1 i_2 \ldots i_r$ into $i_2 \ldots i_r i_1$. The properties of the trace ensure that these two sequences have the same probability, so that, while (9.3.8) no longer holds, (9.3.2) still does.

Notice that we are dealing with a Markov chain with an infinity of states, so that we need to take care to verify that the process converges.

CHAPTER 10

Long Polymer Molecules

10.1 Self-avoiding walks

According to the Mayer model, the properties of a long-chain polymer molecule are crudely but sufficiently well represented if one supposes that the successive atoms occupy adjacent sites of a regular tetrahedral† lattice. If one treats all such configurations as equally probable, the positions of the atoms in a chain of $(n+1)$ atoms are distributed in exactly the same way, relative to the position of one end, as the sites passed through in n steps of a Pólya walk (random walk) on the same lattice. One may then appeal to the theory of Pólya walks for the various information that one requires; of particular interest is the mean square distance $\langle r_n^2 \rangle$ between the ends of the chain.

This treatment has the flaw that it allows more than one atom to occupy the same site, which ought, on account of the mutual repulsive forces, to be physically impossible. This is known as the *excluded volume* effect. One needs, therefore, to modify the theory to consider only *self-avoiding* walks, which, as their name implies, never intersect themselves. Theoretical treatment at once becomes virtually impossible. For instance, it is easy to show that $\langle r_n^2 \rangle$ for a Pólya walk is proportional to n, and to find the constant of proportionality for any particular lattice, but for self-avoiding walks not even the asymptotic form of $\langle r_n^2 \rangle$ for large n is known.

† To describe the tetrahedral lattice simply we need to divide the sites into two classes, *odd* and *even*. Then each even site is adjacent to 4 odd sites, reached from it by the vectors $(1,1,1)$, $(1,-1,-1)$, $(-1, 1,-1)$, and $(-1, -1,1)$, and each odd site is adjacent to 4 even sites, reached by the inverse vectors $(-1,-1,-1)$, $(-1,1,1)$, $(1,-1,1)$, and $(1,1,-1)$.

Almost the only proved fact about self-avoiding walks [1] is that, if $f(n)$ is the number of self-avoiding walks of n steps on a regular lattice, there is a constant k, depending on the lattice, such that

$$0 \leqslant k = \inf_{n \geqslant 1} n^{-1} \log f(n) = \lim_{n \to \infty} n^{-1} \log f(n) < \infty. \dagger$$

(10.1.1)

If the number of unrestricted walks of n steps is $c^n (c = 4$ in the case of the tetrahedral lattice), then $\mu = \log c - k$ is called the *attrition constant* of the lattice, and the proportion of all unrestricted walks that do not intersect themselves in the first n steps is asymptotic to $e^{-\mu n}$. No one has yet succeeded in determining theoretical values for k or μ. For a review of theoretical results, see [3], [4], [5], [6], [7], [8], and [9].

We have here, therefore, two problems to which Monte Carlo methods might provide solutions; estimating k or μ, and either estimating $\langle r_n^2 \rangle$ or (rather more difficult) finding a possible asymptotic formula for $\langle r_n^2 \rangle$ for large values of n. It is true that these problems are unique, so that any method that may be devised to solve them is unlikely to have any more general application; we feel justified, however, in devoting so much attention to them, simply to illustrate once again the vast increases in efficiency that appear when Monte Carlo is applied judiciously rather than blindly. For comments on the practical significance of the results, the reader is referred to [10].

10.2 Crude sampling

We could generate random self-avoiding walks of n steps by the hit-or-miss method of generating random unrestricted walks and rejecting those that intersect themselves before the nth step. This was the method adopted in the earliest attacks on the problem [11], [12]. The proportion of accepted walks is then an unbiased estimator of $e^{\mu n}$, and such walks form a sample from which to estimate $\langle r_n^2 \rangle$. We notice at once that we can save labour by not continuing any walk beyond its first intersection, and by working with several different

† Hammersley and Welsh [2] have shown for a hypercubical lattice that
$$0 \leqslant n^{-1} \log f(n) - k < \gamma n^{-1/2} + n^{-1} \log d \text{ for all } n > n_0(\gamma)$$
where γ is an absolute constant and d is the number of dimensions.

values of n at once; a walk that intersects itself for the first time at the mth step provides instances of n-step self-avoiding walks for all $n < m$.

What concerns us most, however, is to be able to generate some fairly long walks, running to thousands of steps if possible. On account of the attrition, we expect to have to generate a number of the order of $e^{\mu n}$ of walks before we find a self-avoiding walk of n steps. Since μ, for the tetrahedral lattice, is approximately $0\cdot34$, this number is quite large, even for values of n as small as 20. The situation for two-dimensional lattices is worse than for three-dimensional lattices; μ for the square lattice is approximately $0\cdot43$.

We can improve matters slightly by modifying the way in which walks are generated, introducing the condition that no two successive steps are in exactly opposite directions. This has the effect of reducing the attrition μ from $(\log c - k)$ to $(\log(c-1) - k)$, which takes the values $0\cdot06$ and $0\cdot14$ for the tetrahedral and square lattices respectively, while still giving an unbiased sample.

If we go further than this, and modify the procedure to cut out automatically walks that intersect themselves to form a loop of at most $r(\geqslant 2)$ steps, we come into rather deeper waters. Wall, Rubin, and Isaacson [13] build a walk up from previously-constructed self-avoiding *strides* of r steps. It is easy to make up a table to show which of such strides may follow another without intersecting it, and to select successive strides at random from the appropriate rows of this table; however, this procedure gives a biased result unless the number of possible successors of each stride is the same. The simplest way of correcting this is to introduce dummy strides, where necessary, to bring all the numbers of successors up to the same total, with the understanding that when a dummy stride is selected the walk is terminated just as if it had intersected itself.

10.3 Generation of very long walks

All the methods so far discussed still share the defect that a lot of effort is thrown away in short walks, while one is really interested only in long walks. What one wants, therefore, is a scheme to encourage a walk to avoid itself as it is generated. Two such schemes are *inversely restricted sampling* and the *enrichment technique*.

The idea of inversely restricted sampling [14], [15] is that one compels the walk to avoid itself by restricting the choices at each step, and removes the bias with a weighting factor. Suppose that the first n steps of a self-avoiding walk connect the lattice-points $P_0, P_1, \ldots, P_n$. Then the c points that are adjacent to P_n, and so are possible choices for P_{n+1}, each fall into one of three classes: (A) members of the set $P_0, P_1, \ldots, P_n$; (B) points from which there is at least one infinite self-avoiding walk not passing through any point $P_0, P_1, \ldots, P_n$; (C) 'traps' from which it is impossible to continue the self-avoiding walk for more than a finite number of steps. If class B is empty, then P_n must at the previous stage have been in class C.

Supposing for the moment that it is possible to tell immediately to which class each neighbour of P_n belongs (this is in fact the case when one is producing a two-dimensional walk graphically), the technique is then to restrict the choice of P_{n+1} to the class B, at the same time recording the number a_n of members of this class. An unbiased estimator of $f(n)$ is then given by the product $a_1 a_2 \ldots a_n$, or, more precisely, it becomes so if we interpret $f(n)$ as the number of n-step self-avoiding walks *which can be continued indefinitely*, since we have excluded walks that finish up in a trap. Thus a biased estimator of k is given by

$$n^{-1} (\log a_1 + \ldots + \log a_n). \qquad (10.3.1)$$

This bias is caused by the taking of logarithms.

Without graphical aid, things are slightly more difficult, since we cannot distinguish between classes B and C except by trial and error.† The natural policy is to assign every non-member of A in the first instance to class B. If the end-point P_n of the walk is later discovered to belong to class C, we then take back the last step, reduce a_{n-1} accordingly, and try again. The consequence of this procedure is that, when we finally succeed in constructing a walk, the numbers a_i will still overestimate their true values. We may, however, get a sufficiently good estimate of the small correction factor by using the number of

† Marcer [16] has invented an ingenious method on an electronic computer for identifying the traps on the plane square lattice. This will not extend to three dimensions, where, however, traps are much rarer occurrences.

traps that are actually discovered to estimate the number that would have been discovered if every point could have been tested (see [14]).

Having estimated k, one may attempt to establish a formula for $\langle r_n^2 \rangle$ by regression. Possibly the correct form will one day be known, but lacking such knowledge suppose that one makes the plausible assumption

$$\langle r_n^2 \rangle \sim An^\alpha, \quad \text{as } n \to \infty, \tag{10.3.2}$$

where α and A are both unknown and α in particular is to be determined. A suitable method would be to carry out a weighted linear regression of $\log(r_n^2)$ on $\log n$, where a suitable weight would be provided by $(a_1 a_2 \ldots a_n / k^n)$. The data for the regression could be collected by generating several long walks and observing the sub-walks formed by the first 100, 200 ..., steps of each. The great difficulty about this procedure is that the weights are almost certain to get out of hand, a few of them being very much larger than all the rest. This means that the greater part of the data, corresponding to the negligible weights, gets ignored. It is still an open and controversial matter whether or not weighting should be used in this context.

Marcer [16] suggests an alternative estimator of α, making use of the relation

$$\frac{\langle r_n^2 \rangle}{n^{-1} \sum_{j=1}^{n} \langle r_j^2 \rangle} \sim 1 + \alpha \text{ as } n \to \infty \quad (\alpha \geqslant 0). \tag{10.3.3}$$

The enrichment technique [17], [18] is founded on the principle 'hold fast to that which is good' or, if one prefers, on the 'splitting' process discussed in § 8.2. Walks are generated by one of the methods of §10.2, but with the additional rule that, whenever a walk attains a length which is a multiple of s without intersecting itself, m independent attempts are made to continue it. The numbers m and s are fixed, and if $m \approx e^{\mu s}$ the numbers of walks of various lengths generated will be approximately equal. Enrichment has the advantage over inversely restricted sampling that all walks of a given length have equal weight; on the other hand, the dimensions of these walks are liable to be highly correlated (by reason of their having many steps in common), and this should be borne in mind when setting confidence limits to $\langle r_n^2 \rangle$.

Wall and Mazur [19] have made similar calculations allowing for an intermolecular potential in addition to the excluded volume effect. A straightforward sampling or enrichment technique then provides a large number of walks with small probabilities; and, to overcome this effect, Wall, Windwer, and Gans [20], [21] utilized importance sampling.

In connexion with all of these techniques, it is worth the effort to devote some thought to ways of saving time. Unless a graphical method is used, the time taken to generate a self-avoiding walk of n steps is of order n^2 for large n, since the length of the process of testing for intersections is directly proportional to the length of the walk at each step. However, one can exploit the continuity of the walk to make the process quite fast. For example, one could, at every 20th step, say, make a list of all the previously-visited points within 20 steps of the end of the walk; if there is an intersection in the next 20 steps, it can only be with one of these so that there is no need to examine the rest of the walk.

10.4 Walks in continuous space

Fluendy [22] has considered a model rather less artificial than the Mayer model. No longer confined to a lattice, the walk is free to visit any point in space, subject to the restrictions that all steps are of equal length and that every pair of consecutive steps meet at the same angle. The self-avoiding property is reflected in a repulsive potential between every two atoms, similar to that in the gas model of § 9.4, while there is an additional torsion potential depending on the angles between alternate steps such as $P_{n-1}P_n$ and $P_{n-3}P_{n-2}$. Then (compare (9.2.1)) we have

$$\langle r_n^2 \rangle = \frac{\int \ldots \int r_n^2 e^{-\beta E} d\phi_2 \ldots d\phi_n}{\int \ldots \int e^{-\beta E} d\phi_2 \ldots d\phi_n}, \qquad (10.4.1)$$

where $\phi_2, \ldots, \phi_n$ are the successive angles of torsion, r_n is the distance between P_0 and P_n, and E is the total potential

$$E = E_{\text{repulsion}} + E_{\text{torsion}} = \sum_{i > j} \sum E_{ij}^{(R)} + \sum_j E^{(T)}(\phi_j).$$

It would be quite possible to use Metropolis' method (§ 9.3) to estimate $\langle r_n^2 \rangle$. It is more efficient, since in this case it is not difficult, to take part of the exponential factor into the probability density, by the transformation

$$\theta_j = \int \exp(-\beta E^{(T)}(\phi_j)) \, d\phi_j, \qquad (10.4.2)$$

when (10.4.1) becomes

$$\langle r_n^2 \rangle = \frac{\int \ldots \int r_n^2 \exp(-\beta \sum \sum E_{ij}^{(R)}) \, d\theta_2 \ldots d\theta_n}{\int \ldots \int \exp(-\beta \sum \sum E_{ij}^{(R)}) \, d\theta_2 \ldots d\theta_n}. \qquad (10.4.3)$$

In fact, Fluendy obtained quite satisfactory results by crude Monte Carlo estimation of the numerator and denominator of (10.4.3), for walks of up to 12 steps. Long walks are, of course, out of the question on this model.

CHAPTER 11

Percolation Processes

11.1 Introduction

Percolation processes [1] deal with deterministic flow in a random medium, in contrast with *diffusion processes* which are concerned with random flow in a deterministic medium. The terms 'flow' and 'medium' are abstract and bear various interpretations according to context. Frisch and Hammersley [2] discuss these interpretations in the different applications of percolation theory to physical problems.

Although a certain amount of general mathematical theory now relates to percolation processes, this theory is qualitative and, with only a few slight exceptions, Monte Carlo methods provide the only known way of obtaining quantitative answers. It will suffice here to describe a typical percolation problem together with its Monte Carlo solution.

11.2 Bond percolation on the cubic lattice

If a lump of porous material is put in a bucket of water, will the interior of the lump get wet and, if so, to what extent? We visualize the material as a network of interconnecting pores, some of which are large enough to convey water and others so small that they block its passage. We shall idealize this situation by supposing that the structure of the pores forms a simple cubic lattice. More precisely the places where the pores interconnect are called *sites*, and there is one site at each point with integer co-ordinates (x, y, z) in three-dimensional Euclidean space. Two sites are called *neighbours* if they are unit distance apart; and we suppose that each pair of neighbouring sites is connected by a pore (which we shall call a *bond* for the sake of conformity with the nomenclature in other applications of percolation processes). Each bond, independently of all other bonds, has a prescribed probability p (the same for each bond) of being large enough to transmit water, and a probability $q = 1 - p$ of being too small and therefore unable to transmit water. We call the bonds *blocked* or *unblocked*, as the case may be. An unblocked bond will transmit water in either direction. When the water reaches a site at one end of an unblocked bond, it will travel down any other unblocked bond ending at this site. We imagine the lump of material to be a large chunk (say a cube with M sites along each edge) hewn from the infinite cubic lattice. When the chunk is immersed, all sites on its surface become wet; and from these sites the water flows along unblocked bonds from site to site into the interior. We write $P(p)$ for the proportion of interior sites which become wet when M is very large. We call $P(p)$ the *percolation probability*. It is clearly a non-decreasing function of p, satisfying $P(0) = 0$ and $P(1) = 1$. Percolation theory [1] also tells us that there exists a number p_0, called the *critical probability*, such that $P(p) = 0$ when $0 \leqslant p < p_0$, while $P(p) > 0$ for $p_0 < p \leqslant 1$. Physically, when the proportion of unblocked pores is less than p_0, the water only wets the skin of the lump; but as soon as the proportion exceeds p_0, the water suffuses more or less uniformly throughout the interior of the material.

The problem is to calculate $P(p)$ as a function of p, and hence to determine the critical probability p_0. The problem is probabilistic in its original physical formulation; and, in principle, it should be amenable to direct simulation without any recourse to sophisticated Monte Carlo devices. But this line of attack runs into a prohibitive amount of computing. The simple-minded direct simulation would use random numbers to label each bond as either blocked with probability q or unblocked with probability p, would then examine each dry site in turn, wet it whenever it was connected to an already wet site by an unblocked bond, continue the examination until no more dry sites could be made wet, and finally count the number of wet sites. But if M is at all large, say $M = 200$, there are 8 million sites and 24 million bonds in the chunk. This calls for enormous storage facilities in the computer, even if we go to the trouble of representing the state of each bond (blocked or unblocked) and each site (dry or wet) by a single binary digit and packing this information into separate bits of the various stored words. Further, we have to scan all

8 million sites over and over again until no more can be wet, and this may require one or two hundred repetitions of the scan. Moreover, this procedure only affords a single experimental observation on the required number $P(p)$; so that we must repeat it several times to ensure that sampling errors are duly accounted for. Lastly, what we have so far described relates merely to one prescribed value of p; so that we must repeat the work for, say, 50 different values of p in order to get a graph of $P(p)$ against p for $0 \leq p \leq 1$. All in all, the complete calculation would need to process about 10^{12} or 10^{13} pieces of information and would keep a modern high-speed computer continuously busy for about 50 years [3]. Direct simulation is thus out of the question.

We shall surmount this difficulty in two stages. In the first stage, we look for another probabilistic problem, which also has $P(p)$ for its solution. As explained in § 1.1, we use mathematical theory to connect these two problems. In the first problem (the original one) we started the water at *all* sites of the surface of the chunk and followed its inward flow. In the second problem, we shall start the water at just *one* fixed interior site (to be called the *source* site) and follow its outward flow from there. We shall write $P_N(p)$ for the probability that, in this second problem, the water will ultimately wet at least N other sites. The theory of percolation processes [1] tells us that (under mild conditions here satisfied $P_N(p) \to P(p)$ as $N \to \infty$. Hence by estimating $P_N(p)$ for some large value of N, we can estimate the required function $P(p)$. It turns out that $N \sim 6000$ is a sufficiently large number.

The first consequence of this change of viewpoint is that we enormously reduce the storage requirements. For now we have only got to continue the Monte Carlo experiment until either (i) we have succeeded in wetting N sites or (ii) fewer than N sites are wet and no unblocked path leads from any of *these* sites to a dry atom. We can wet the sites one at a time, starting from the original single source site, and we have merely to look at the bonds leading from the so-far-wetted sites to see if the process of further wetting can be continued. At most we need only store information relating to N sites and the particular bonds (at most $5N+1$ of them) leading from these N sites. Thus the total storage requirement is now only about $6N$ instead of M^3. Moreover, we also reduce the number of pieces of information to be handled by the computer from a multiple of M^4 to a multiple of N^2 (N^2 rather than N arises because we must successively sort partial lists of N sites). In the present instance, the storage requirements come within the realm of practical computing and the total computing time comes down from about 50 years to something like 1 year. Of course, this is still a prohibitive amount of computing; but the reduction of a factor of 50 or so is still quite remarkable when one considers the very simple change of viewpoint (outward flow instead of inward) which achieves it.

To reduce the computing time still further to manageable proportions, we embark on a second equally simple transformation of the problem. In our second version of the problem we were still faced with the labour of repeating the Monte Carlo experiment for various values of p in order to build up a graph of $P(p)$. For the third version of the problem, we adopt a device which allows us to calculate $P(p)$ simultaneously for all values of p from one and the same Monte Carlo experiment.

In the first and second versions each bond was either blocked or unblocked. In the third version we shall assign a rectangularly distributed random variable ξ independently to each bond. Instead of starting a single fluid (water) from the single source site, we shall simultaneously start infinitely many different fluids all from the same single source site. Each of the separate fluids will be characterized by a number g: in fact, there will be one fluid for each number g in $0 \leqslant g \leqslant 1$, and we shall call the fluid, characterized by the number g, the g-fluid. We adopt the rule that a particular bond with an assigned random ξ will be capable of transmitting the g-fluid if and only if $g \leqslant \xi$. The consequence of this rule is that, if the g_0-fluid succeeds in wetting a particular set of sites S, then all g-fluids, for which $g < g_0$, will also wet S. Let us write g_N for the maximum value of g such that g-fluids wet N or more sites. We now prove that

$$P(g_N \geqslant 1-p) = P_N(p). \qquad (11.2.1)$$

To see the truth of (11.2.1) let us take any fixed value of p, and consider what happens to the particular $(1-p)$-fluid. This fluid will traverse any particular bond, to which the random number ξ has been assigned, if and only if $1-p \leqslant \xi$; and the probability of this event is

$P(\xi \geqslant 1-p) = p$ since ξ is rectangularly distributed. Thus the $(1-p)$-fluid in the third version of the problem behaves exactly like the single fluid in the second version of the problem. If in the second version the fluid succeeds in wetting at least N sites, then in the third version the $(1-p)$-fluid will also do so; and therefore a *a fortiori* $g_N \geqslant 1-p$. On the other hand if $g_N < 1-p$, the $(1-p)$-fluid will not wet as many sites as N and neither will the single fluid in the second version of the problem. The last two sentences are equivalent to the equation (11.2.1), which is consequently established. From (11.2.1) we deduce

$$P(1-g_n \leqslant p) = P_N(p); \qquad (11.2.2)$$

and this asserts that $P_N(p)$ is the distribution function of the random variable $1-g_N$. Hence to solve our original problem, it suffices to find the distribution function of $1-g_N$ in the third version of the problem. Let us see how to do this.

We define $g_0^* = 1$ and S_0 to be the single source site. We now calculate recursively for $n = 0, 1, \ldots$ the quantities g_n^* and a sequence of sets of sites, the nth set being denoted by S_n. (We shall prove presently that $g_N^* = g_N$; so that the recursive calculation will actually yield the wanted variable g_N.) Suppose that we have already determined g_n^* and S_n and have shown that S_n consists of the source site and n other sites, each of which is wet by the g_n^*-fluid. By definition, this is certainly true for $n = 0$. We now examine all bonds which lead from a site of S_n to a site not in S_n; and we take from amongst them that bond which has the greatest value of ξ assigned to it. Let η_n denote this maximum ξ, and let A_n be the site lying at the far end of this bond (i.e. not belonging to S_n). We then take S_{n+1} to be the union of S_n and A_n; and we define g_{n+1}^* to be the smaller of η_n and g_n^*. This construction validates the recursion, for clearly S_{n+1} consists of the source site and $n+1$ other sites, each of which will be wet by the g_{n+1}^*-fluid. Thus the g_n^*-fluid wets n sites besides the source site; and therefore

$$g_n^* \leqslant g_n, \qquad (11.2.3)$$

by the definition of g_n as the maximum characterizing number with this property. On the other hand, we chose η_n to have the maximum value for all bonds leading out of S_n; and thus no g-fluid with $g > \eta_n$

can escape from S_n. Since $S_0, S_1, \ldots$ are successively contained within each other, by their construction, it follows that no g-fluid, starting at the source site, can reach any site outside S_n if

$$g > \min(\eta_0, \eta_1, \ldots \eta_n) = g_{n+1}^*. \tag{11.2.4}$$

However, any set of $n+1$ sites, which do not include the source site, must have at least one site outside S_n. Thus (11.2.4) shows that

$$g_{n+1} \leqslant g_{n+1}^*. \tag{11.2.5}$$

Because (11.2.3) and (11.2.5) hold for all n, we conclude that

$$g_n = g_n^*; \tag{11.2.6}$$

and we can drop the asterisks on g_n^*.

To get a sample of values of g_N, we may repeat the recursive calculation k times, say, assigning fresh random ξ's to the bonds at each repetition. This will give a sample of k values of g_N; and the proportion of these values satisfying $g_N \geqslant 1 - p$ will be an estimate of $P_N(p)$ for each p in $0 \leqslant p \leqslant 1$.

The actual method of performing the recursive calculation of g_n on a computer is a fairly technical piece of programming, which need not concern us here but is described more fully in [3] and [4].

11.3 Refinement of the Monte Carlo method by use of ratio estimates

Let us define

$$g_{mN} = \min(\eta_{m+1}, \eta_{m+2}, \ldots, \eta_N), \tag{11.3.1}$$

where the η's are those defined in § 11.2. Then (11.2.4) yields

$$g_N = \min(g_m, g_{mN}); \tag{11.3.2}$$

and we have

$$P(g_N \geqslant 1 - p) = P(g_m \geqslant 1 - p) P(g_{mN} \geqslant 1 - p | g_m \geqslant 1 - p). \tag{11.3.3}$$

We may write this as

$$P_N(p) = P_m(p) P_{mN}(p), \tag{11.3.4}$$

where $P_{mN}(p)$ is the conditional probability

$$P_{mN}(p) = P(g_{mN} \geqslant 1 - p | g_m \geqslant 1 - p). \tag{11.3.5}$$

The Monte Carlo recursive calculation described in § 11.2 provides g_n for $n = 1, 2, \ldots, N$: and hence yields an estimate $\hat{P}_n(p)$ of $P_n(p)$ for $n = 1, 2, \ldots, N$. It does not provide a direct estimate $\hat{P}_{mN}(p)$ of $P_{mN}(p)$; but we could obtain one from the equation

$$\hat{P}_N(p) = \hat{P}_m(p)\,\hat{P}_{mN}(p), \quad (N > m). \tag{11.3.6}$$

Now most of the sampling variations in g_N arise from variations in the first few terms in $\eta_0, \eta_1, \ldots$, since the early terms depend upon the assigned ξ's of only a few bonds, while the later terms depend upon more ξ's and are therefore more stable. That is to say, most of the sampling error in $\hat{P}_N(p)$ in (11.3.6) depends upon the sampling errors in $\hat{P}_m(p)$ if m is small. We should get a more precise estimate $P_N^*(p)$ of $P_N(p)$ if we replaced $\hat{P}_m(p)$ by its true value $P_m(p)$:

$$P_N^*(p) = P_m(p)\,\hat{P}_{mN}(p), \quad (N > m). \tag{11.3.7}$$

From (11.3.6) and (11.3.7) we get the ratio estimator

$$P_N^*(p) = [P_m(p)/\hat{P}_m(p)]\,\hat{P}_N(p), \quad (N > m), \tag{11.3.8}$$

which we can use if $P_m(p)$ is known theoretically. For very small m it is possible to work out $P_m(p)$ exactly: thus

$$P_1(p) = 1 - q^6, \quad P_2(p) = 1 - q^6 - 6pq^{10}, \tag{11.3.9}$$

where $q = 1 - p$. For $m > 2$, it becomes increasingly laborious to calculate $P_m(p)$; but we can nevertheless usefully employ (11.3.8) with $m = 2$. Indeed, as an illustrative check of this procedure, consider what happens when $m = 1$, and $N = 2$. The Monte Carlo experiment [3] yielded $\hat{P}_1(0\cdot30) = 0\cdot844$ and $\hat{P}_2(0\cdot30) = 0\cdot789$. From (11.3.9). $P_1(0\cdot30) = 0\cdot882$; and hence (11.3.8) gives $P_2^*(0\cdot30) = 0\cdot825$. The true value is $P_2(0\cdot30) = 0\cdot831$; and we see that P_2^* is a better estimate of P_2 than $\hat{P}_2$ is.

This use of ratio estimators is a typical example of the way in which Monte Carlo results may be improved by replacing a part of the experiment by exact analysis.

11.4 Other Monte Carlo approaches to percolation problems

There are similar percolation problems, in which the blockages occur randomly at the sites instead of on the bonds: and the percolation may occur on types of lattice other than the simple cubic. The calculation is essentially the same: for details and numerical results see [2], [4], [5], [6] and [7].

A different kind of technique, based on the sizes of clusters of sites which may be connected together by unblocked bonds (or which, in the site problem, consist of neighbouring unblocked sites), is given by Dean [8]. This yields estimates of the critical probability p_0; but does not provide direct information on $P(p)$.

CHAPTER 12

Multivariable Problems

12.1 Introduction

As Thacher [1] has remarked, 'one of the areas of numerical analysis that is most backward and, at the same time, most important from the standpoint of practical applications' is multivariable analysis. Much of our ignorance results from a lack of adequate mathematical theory in this field. Unless a function of a large number of variables is pathologically irregular, its main features will depend only upon a few overall characteristics of its variables, such as their mean value or variance, but we have little, if any, theory to tell us when this situation appertains and, if so, what overall characteristic it is that actually describes the function. In these circumstances an empirical examination, effectively some sampling experiment on the behaviour of the function, may guide us.

The Monte Carlo methods hitherto discussed mostly turn upon sampling from a population of random numbers. In this chapter we shall consider sampling from the population of terms in a mathematical formula. While this distinction is worth making in the hope that it may stimulate fresh avenues of thought, it is in no way a clear-cut distinction; as with most Monte Carlo work, if not most of mathematics, one formulation of a problem may be recast in terms of another. For example, an integral (with a continuous integrand) can be expressed, to within as small an error as we please, as the sum of a large number of terms making up a weighted sum of the values of the integrand, and the ordinary Monte Carlo estimation of the integral could be regarded as a sampling from amongst such terms. Again, the second half of § 12.3 is only a particular case of § 7.5, and perhaps a bad case at that.

The material of this chapter is largely speculative, but on that account perhaps offers good scope for further research. We only consider a few rather disconnected topics, chosen mainly because they happened to be lying around. It is a postulate that, underlying each of them, there is some kind of uniformity or regularity attached to the function of many variables.

Some other multivariable problems, which we only mention here and pursue no further, are the many-body problem of celestial mechanics, numerical weather prediction and similar complicated hydrodynamic or magnetohydrodynamic problems, the distribution of eigenvalues of a large vibrating system (as in solid-state physics), and large-scale econometric problems including linear and dynamic programming. Any of these topics offer challenging Monte Carlo problems for the bold research worker.

12.2 Interpolation of functions of many variables

Consider a function $f(x) = f(x_1, x_2, \ldots, x_n)$ of a large number of variables, say $n = 1000$. Suppose that we have an algorithm which enables us to calculate f at any vertex of the unit cube, that is to say when each x_i is either 0 or 1. The problem is to interpolate f at some given point $\mathbf{p} = (p_1, \ldots, p_n)$ within this cube $(0 \leqslant p_i \leqslant 1)$; we shall suppose for simplicity that linear interpolation is adequate. In principle, the problem is elementary: we have only to compute

$$f(\mathbf{p}) = \sum r_1 r_2 \ldots r_n f(\delta_1, \delta_2, \ldots, \delta_n), \qquad (12.2.1)$$

where the sum is taken over all combinations of $\delta_i = 0$ or 1, and where $r_i = 1 - p_i$ or p_i according as $\delta_i = 0$ or 1. In practice, however, this procedure is unworkable, because the right-hand side of (12.2.1) is a sum of 2^n terms, each term being a product of $(n+1)$ quantities. Even if we had time enough for the calculation, the loss of accuracy due to rounding errors in such a large number of very small terms would invalidate the answer.

As a substitute procedure, we may sample† terms from the formula (12.2.1). Choose a set of numbers

$$\eta_i = \begin{cases} 1 \text{ with probability } 1 - p_i \\ 0 \text{ with probability } p_i \end{cases} \qquad (12.2.2)$$

† It would be particularly interesting to know what happens if we sample using quasirandom numbers.

independently for $i = 1, 2, \ldots, n$. Then

$$P(\eta_1 = \delta_1, \eta_2 = \delta_2, \ldots, \eta_n = \delta_n) = r_1 r_2 \ldots r_n, \quad (12.2.3)$$

and hence, by (12.2.1),

$$t = f(\eta_1, \eta_2, \ldots, \eta_n) \quad (12.2.4)$$

is an unbiased estimator of $f(\mathbf{p})$. More profitably, we may take the mean of N such values of (12.2.4) as an estimator of $f(\mathbf{p})$; this will have a standard error

$$\sigma = [N^{-1} \operatorname{var} f(\eta_1, \ldots, \eta_n)]^{1/2}. \quad (12.2.5)$$

Is σ small enough for this to be a reasonable procedure for moderate values of N? The answer depends, naturally enough, upon how well-behaved f is. We may describe the regularity of f in terms of the difference $\Delta_i(V)$, where $\Delta_i(V)$ denote the difference between f at the vertex V of the unit cube and at the ith one of the n vertices which are nearest to V. Let $M(r)$ denote the smallest function of r such that

$$\sum_{(r)} [\Delta_i(V)]^2 \leqslant M(r) \quad (12.2.6)$$

holds for each vertex V of the unit cube, where $\sum_{(r)}$ denotes summation over any subset of r of the n possible values of i. Then Hammersley [2] proved that

$$\sigma \leqslant N^{-1/2} [\tfrac{1}{4} M(1) + \tfrac{1}{2} \sum_{r=2}^{n} M(r)/r]^{1/2}. \quad (12.2.7)$$

There are two special but important cases of (12.2.7). The first case arises when all the partial derivatives of f are bounded and we may write $M(r) = Kr$, where K is a constant. This gives

$$\sigma < \sqrt{(Kn/2N)}, \quad (12.2.8)$$

so that we can get a reasonable estimate of the interpolate by making N an adequate multiple of n. Even with $n = 1000$, we can probably handle the appropriate value of N on a high-speed computer. The second case arises when $M(r)$ is independent of r. This occurs if (the finite difference analogue of) grad f is uniformly bounded. Then

$$\sigma < \sqrt{[(\tfrac{1}{2}+\log n)\,K/2N]},\qquad(12.2.9)$$

where $K \geqslant M(r)$ is the constant that bounds $M(r)$. Since the right-hand side of (12.2.9) is a very slowly increasing function of n, we can envisage work with prodigiously large value of n, such as $n = 10^{26}$. However, we now run into other difficulties; with very large n, it will be no trivial matter to select a vertex by means of (12.2.2), which involves drawing n random η_i. Unless there are other conditions on f, such as a high degree of symmetry between co-ordinates, the method will be unworkable.

Thus far, we have considered only linear interpolation. The prospects for higher-order interpolation do not appear good. Once more, the procedure is simple in principle. We have the multi-Lagrangian formula

$$f(x_1,\ldots,x_n) = \sum_j f(x_{1j},\ldots,x_{nj}) \prod_{i,k}{}' \frac{(x_i-x_{ik})}{(x_{ij}-x_{ik})}\qquad(12.2.10)$$

where j ranges over all the data points x_{ij}, and where $\prod'$ denotes the product omitting terms of the type $(x_i-x_{ij})/(x_{ij}-x_{ij})$. We may write (12.2.10) more concisely as

$$f = \sum_j L_j f_j = \sum_j p_j(L_j f_j/p_j).\qquad(12.2.11)$$

Thus, if $p_j > 0$ and $\sum_j p_j = 1$, we choose the jth datum point with probability p_j and use $L_j f_j/p_j$ as an unbiased estimate of f. The p_j play the usual part of importance sampling, so that we may hope to select them in a way which simplifies the calculation of $L_j f_j/p_j$ and reduces its standard error.

In practice, however, it seems difficult to achieve these ends. We still have the identity

$$\sum_j L_j = 1,\qquad(12.2.12)$$

as may be deduced from (12.2.11) by taking $f \equiv 1$; but, when f is relatively flat and we would like to take p_j more or less proportional to L_j, we are prevented by the fact that some of the L_j are negative. Indeed this change of sign inflates the standard error, for we have

$$
\begin{aligned}
\mathrm{var}\, f &= \sum_j p_j \,(L_j f_j / p_j)^2 - \left(\sum_j L_j f_j\right)^2 \\
&= \left[\sum_j \left(\frac{|L_j f_j|}{\sqrt{p_j}}\right)^2\right]\left[\sum_j (\sqrt{p_j})^2\right] - \left(\sum_j L_j f_j\right)^2 \\
&\geqslant \left(\sum_j |L_j f_j|\right)^2 - \left(\sum_j L_j f_j\right)^2,
\end{aligned} \tag{12.2.13}
$$

by Cauchy's inequality. The right-hand side of (12.2.13) presents a minimum variance which we cannot improve upon however ingeniously we sample. Further the right-hand side of (12.2.13) may easily be very large in comparison with f. To take a simple illustration, suppose that $f \equiv 1$ and that the data points consist of all possible combinations of the form $(x_1', x_2', \ldots, x_n')$, where x_i' belongs to a given set of points Λ_i. If $L_j^{(i)}$ denotes the univariate Lagrangian interpolation coefficients for Λ_i, we have

$$
\sum_j |L_j| = \prod_{i=1}^n \left(\sum_j |L_j^{(i)}|\right). \tag{12.2.14}
$$

If Λ_i consists of more than two points, $\sum_j |L_j^{(i)}| > 1$; so (12.2.14) and hence $\mathrm{var}\, f$ can be exponentially increasing functions of n.

The matter is discussed at greater length in [2], but the general conclusion is that non-linear interpolation of a function of many variables presents a completely unsolved problem. A possible way round the difficulty is to stratify the sample.

12.3 Calculations with large matrices

Suppose we wish to calculate the element c_{ij} of a matrix product $\mathbf{C} = \mathbf{AB}$, where $\mathbf{A} = (a_{ij})$ and $\mathbf{B} = (b_{ij})$ are matrices of large order $l \times m$ and $m \times n$. We know that

$$
c_{ij} = \sum_{k=1}^m a_{ik} b_{kj}. \tag{12.3.1}
$$

If the right-hand side of (12.3.1) contains an unmanageably large number of terms, we may select N of them uniformly at random. Then

$$\gamma_{ij} = \frac{m}{N} \sum_{r=1}^{N} a_{ik_r} b_{k_r j} \qquad (12.3.2)$$

will be an unbiased estimator of c_{ij} with standard error

$$\left[\frac{(m-N)}{N(m-1)}\right]^{1/2}\left[m \sum_{k=1}^{m} a_{ik}^2 b_{kj}^2 - \left(\sum_{k=1}^{m} a_{ik} b_{kj}\right)^2\right]^{1/2}. \qquad (12.3.3)$$

The second factor in (12.3.3) will be roughly of the same order of magnitude as c_{ij} if the products $a_{ik} b_{kj}$ are mostly of the same approximate size. Thus if $0 < \lambda \leqslant a_{ik} b_{kj} \leqslant \Lambda$ for $k = 1, \ldots, m$, we have

$$\left[m \sum_{k=1}^{m} a_{ik}^2 b_{kj}^2 - \left(\sum_{k=1}^{m} a_{ik} b_{kj}\right)^2\right]^{1/2} \leqslant c_{ij}(\Lambda-\lambda)/2\sqrt{(\Lambda\lambda)}.$$
$$(12.3.4)$$

So (12.3.2) will have a fairly acceptable standard error if $N \sim 10^3$; and the Monte Carlo method will be worth considering if $m \sim 10^5$, say.

This technique can be applied to the estimation of eigenvectors of a large matrix $\mathbf{A}$. We recall that the sequence of vectors, starting from an arbitrary vector $\mathbf{u}_1$ and continuing via

$$\mathbf{u}_{n+1} = \mathbf{A}\mathbf{u}_n, \qquad (12.3.5)$$

is such that

$$\mathbf{x}_N = N^{-1} \sum_{n=1}^{N} \mathbf{u}_n/||\mathbf{u}_n|| \qquad (12.3.6)$$

converges as $N \to \infty$ to the eigenvector $\mathbf{x}$ of $\mathbf{A}$: here $||\mathbf{u}||$ is a norm of $\mathbf{u}$, say the square root of the sum of the squares of its elements. To simulate this by Monte Carlo, we may select m sequences ($i = 1, 2, \ldots, m$) of random integers $k_i(1)$, $k_i(2)$, ... independently and uniformly distributed over the integers $1, 2, \ldots, m$, and from an arbitrary vector $\mathbf{v}_1$ compute $\mathbf{v}_n = \{v_{n1}, \ldots, v_{nm}\}$ via

$$v_{n+1, i} = \alpha_n a_{ik_i(n)} v_{nk_i(n)} \quad (i = 1, \ldots, m), \qquad (12.3.7)$$

where α_n is chosen such that $||\mathbf{v}_{n+1}|| = 1$. On these we base the estimator

$$N^{-1} \sum_{n=1}^{N} \mathbf{v}_n \qquad (12.3.8)$$

which will be a slightly biased estimator of $\mathbf{x}_N$. The bias arises because of the normalizing factor α_n, but will not be serious if m is large.

However, so far as we are aware, it is not yet known how well these proposals work in practice, if indeed they work at all.

12.4 Reduction of multivariable functions

A multivariable function is more easily handled if it can be expressed to an adequate degree of approximation in the form

$$f(x_1,\ldots,x_n) = \sum_{r=1}^{R} g_{r1}(x_1)\,g_{r2}(x_2)\ldots g_{rn}(x_n). \qquad (12.4.1)$$

Allen [3] has shown how to compute such a representation when $n = 2$, and his treatment was extended to general values of n in [2]. The relevant formulae involve various tensor sums with a great many terms in them. The only hope of calculating them seems to be through some method which samples only a small fraction of terms.

12.5 Network problems

Recently there has been much interest in the difficult mathematical problems connected with networks or linear graphs. A network consists of a number of nodes, certain pairs of which are connected together by arcs. There may be numerical or non-numerical functions associated with the nodes and/or with the arcs.

The percolation processes discussed in Chapter 11 are examples of network problems. Another example arises in a system of theorems in logic. Each node is a theorem, that is to say a logical proposition, while each arc is a logical operation. The arc from node M to node N carries the operation which, when applied to the proposition M yields the proposition N. A subset of the nodes represents postulates. The problem is then to discover whether or not a given proposition can be proved from these postulates, that is to say whether this node can be reached from the postulate-nodes via arcs of the network.

If it can be so reached, which is the shortest path (i.e. the 'neatest' proof)? In industrial production, the arcs may represent machining operations, and path along such arcs from node to node will represent a succession of operations applied to a mechanical component. Each arc may carry a number indicating, say, the time taken to perform the operation. We may then be interested in the path of shortest time through a specified set of nodes, possibly in a certain order. There are many other kinds of variation of this problem [4].

Usually there are many nodes and many possible paths, so many that a complete enumeration of the situation is impossible. This suggests a fruitful field for sampling and search procedures, but as yet little Monte Carlo work has been done here. There are challenging problems here for research into Monte Carlo techniques on multivariable problems.

References

Chapter 1

1 P. H. LESLIE and D. CHITTY (1951). 'The estimation of population parameters from data obtained by means of the capture-recapture method. I. The maximum likelihood equations for estimating the death rate.' *Biometrika*, **38**, 269–292.

2 W. E. THOMSON (1957). 'Analogue computers for stochastic processes.' *Abbreviated Proceedings of the Oxford Mathematical Conference for Schoolteachers and Industrialists*, 54–57. London: The Times Publishing Company.

3 A. W. MARSHALL (1956). 'An introductory note.' *Symposium on Monte Carlo methods*, ed. H. A. MEYER, 1–14. New York: Wiley.

4 A. HALL (1873). 'On an experimental determination of π' *Messeng. Math.* **2**, 113–114.

5 J. H. CURTISS *et al.* (1951). 'Monte Carlo method.' *National Bureau of Standards Applied Mathematics Series*, **12**.

6 Lord KELVIN (1901). 'Nineteenth century clouds over the dynamical theory of heat and light.' *Phil. Mag.* (6) **2**, 1–40.

7 K. W. MORTON (1956). 'On the treatment of Monte Carlo methods in textbooks.' *Math. Tab. Aids Comput.* **10**, 223–224.

Chapter 2

1 W. G. COCHRAN (1953). *Sampling techniques.* New York: Wiley.

2 H. CRAMÉR (1946). *Mathematical methods of statistics.* Princeton Univ. Press.

3 M. G. KENDALL and A. STUART (1961). *The advanced theory of statistics*, Vols I–III. London: Charles Griffin.

4 R. L. PLACKETT (1960). *Principles of regression analysis.* Oxford: Clarendon Press.

Chapter 3

1 M. G. KENDALL and B. BABINGTON SMITH (1939). *Tables of random sampling numbers.* Tracts for Computers, **24**. Cambridge University Press.

2 RAND CORPORATION (1955). *A million random digits with 100,000 normal deviates.* Glencoe, Illinois: Free Press.

3 Lord KELVIN (see chapter 1, reference 6).

4 T. E. HULL and A. R. DOBELL (1962). 'Random number generators.' *Soc. Indust. Appl. Math. Rev.* **4**, 230–254.

5 G. E. FORSYTHE (1951). 'Generation and testing of random digits.' *National Bureau of Standards Applied Mathematics Series*, 12, 34–35.

6 D. H. LEHMER (1951). 'Mathematical methods in large-scale computing units.' *Ann. Comp. Lab. Harvard Univ.* **26**, 141–146.

7 M. GREENBERGER (1961). 'Notes on a new pseudorandom number generator.' *J. Assoc. Comp. Mach.* **8**, 163–167.

8 M. GREENBERGER (1961 and 1962). 'An *a priori* determination of serial correlation in computer generated random numbers.' *Math. Comp.* **15**, 383–389; and corrigenda *Math. Comp.* **16**, 126.

9 O. TAUSSKY and J. TODD (1956). 'Generation and testing of pseudo-random numbers.' *Symposium on Monte Carlo methods*, ed. H. A. MEYER, 15–28. New York: Wiley.

10 I. J. GOOD (1953). 'The serial test for sampling numbers and other tests for randomness.' *Proc. Camb. phil. Soc.* **49**, 276–284.

11 K. F. ROTH (1954). 'On irregularities of distribution.' *Mathematika*, **1**, 73–79.

12 J. C. VAN DER CORPUT (1935). 'Verteilungsfunktionen.' *Proc. Kon. Akad. Wet. Amsterdam*, **38**, 813–821, 1058–1066.

13 J. H. HALTON (1960). 'On the efficiency of certain quasi-random sequences of points in evaluating multidimensional integrals.' *Numerische Math.* **2**, 84–90 and corrigenda p. 196.

14 R. D. RICHTMYER (1958). 'A non-random sampling method, based on congruences, for Monte Carlo problems.' *Inst. Math. Sci. New York Univ. Report* NYO-8674 *Physics*.

15 C. B. HASELGROVE (1961). 'A method for numerical integration.' *Math. Comp.* **15**, 323–337.

16 J. VON NEUMANN (1951). 'Various techniques used in connection with random digits.' *National Bureau of Standards Applied Mathematics Series*, **12**, 36–38.

17 J. W. BUTLER (1956). 'Machine sampling from given probability distributions.' *Symposium on Monte Carlo methods*. ed. H. A. MEYER. 249–264. New York: Wiley.

18 G. MARSAGLIA (1961). 'Generating exponential random variables.' *Ann. Math. Statist.* **32**, 899–900.

19 G. MARSAGLIA (1961). 'Expressing a random variable in terms of uniform random variables.' *Ann. Math. Statist.* **32**, 894–898.

20 D. TEICHROEW (1953). 'Distribution sampling with high-speed computers.' *Thesis*. Univ. of North Carolina.

21 M. E. MULLER (1959). 'A comparison of methods for generating normal deviates.' *J. Assoc. Comp. Mach.* **6**, 376–383.

22 G. E. P. BOX and M. E. MULLER (1958). 'A note on the generation of random normal deviates.' *Ann. Math. Statist.* **29**, 610–611.

23   J. C. BUTCHER (1961). 'Random sampling from the normal distribution.' *Computer J.* **3**, 251–253.

24   H. WOLD (1948). *Random normal deviates.* Tracts for computers, **25**. Cambridge Univ. Press.

25   FRANKLIN, J. N. (1963). 'Deterministic simulation of random processes.' *Math. Comp.* **17**, 28–59.

26   D. G. CHAMPERNOWNE (1933). 'The construction of decimals normal in the scale of ten.' *J. London Math. Soc.* **8**, 254–260.

27   A. H. COPELAND and P. ERDÖS (1946). 'Note on normal numbers.' *Bull. Amer. Math. Soc.* **52**, 857–860.

Chapter 4

1   ANONYMOUS (1957). 'Calculating machine. Electronics applied to water storage on Nile.' *The Times* (22 March 1957) 9. London.

2   P. A. P. MORAN (1959). *The theory of storage.* London: Methuen.

3   K. D. TOCHER (1961). *Handbook of the general simulation program.* I. United Steel Co. Dept. Oper. Res. Cyb. Rep. 77/ORC 3/TECH.

4   K. D. TOCHER (1963). *The art of simulation.* London: English Universities Press.

5   M. S. BARTLETT (1960). *Stochastic population models in ecology and epidemiology.* London: Methuen.

6   Colonel A. W. DEQUOY (1962). 'U.S. Army Strategy and Tactics Analysis Group. The U.S. Army's War Gaming Organization.' *Proc. 7th Conference on the Design of Experiments in Army Research, Development, and Testing*, 585–592. Durham: U.S. Army Research Office.

7   D. G. MALCOLM (1960). 'Bibliography on the use of simulation in management analysis.' *J. Oper. Res. Soc. America*, **8**, 169–177.

8   J. M. HAMMERSLEY (1961). 'On the statistical loss of long-period comets from the solar system. II' *Proc. 4th Berkeley Symposium on Mathematical Statistics and Probability*, **3**, 17–78.

9   J. K. MACKENZIE and M. J. THOMSON (1957). 'Some statistics associated with the random disorientation of cubes.' *Biometrika*, **44**, 205–210.

10   D. C. HANDSCOMB (1958). 'On the random disorientation of two cubes.' *Canad. J. Math.* **10**, 85–88.

11   J. K. MACKENZIE (1958). 'Second paper on statistics associated with the random disorientation of cubes.' *Biometrika*, **45**, 229–240.

12   J. M. HAMMERSLEY and J. A. NELDER (1955). 'Sampling from an isotropic Gaussian process.' *Proc. Camb. phil. Soc.* **51**, 652–662.

13   T. DALENIUS, J. HAJEK and S. ZUBRZYCKI (1961). 'On plane sampling and related geometrical problems.' *Proc. 4th Berkeley Symposium on Mathematical Statistics and Probability*, **1**, 125–150.

14   J. E. BEARDWOOD, J. H. HALTON and J. M. HAMMERSLEY (1959). 'The shortest path through many points.' *Proc. Camb. phil. Soc.* **55**, 299–327.

Chapter 5

1 M. G. KENDALL and B. BABINGTON SMITH (see chapter 3, reference 1).
2 T. DALENIUS and J. L. HODGES (1957). 'The choice of stratification points.' *Skandinavisk Aktuarietidskrift*, **3–4**, 198–203.
3 E. C. FIELLER and H. O. HARTLEY (1954). 'Sampling with control variables.' *Biometrika*, **41**, 494–501.
4 J. W. TUKEY (1957). 'Antithesis or regression?' *Proc. Camb. phil. Soc.* **53**, 923–924.
5 J. M. HAMMERSLEY and K. W. MORTON (1956). 'A new Monte Carlo technique: antithetic variates.' *Proc. Camb. phil. Soc.* **52**, 449–475.
6 J. M. HAMMERSLEY and J. G. MAULDON (1956). 'General principles of antithetic variates.' *Proc. Camb. phil. Soc.* **52**, 476–481.
7 D. C. HANDSCOMB (1958). 'Proof of the antithetic variates theorem for $n > 2$.' *Proc. Camb. phil. Soc.* **54**, 300–301.
8 G. H. HARDY, J. E. LITTLEWOOD and G. PÓLYA (1934). *Inequalities.* Cambridge Univ. Press.
9 C. B. HASELGROVE (see chapter 3, reference 15).
10 J. H. HALTON and D. C. HANDSCOMB (1957). 'A method for increasing the efficiency of Monte Carlo integration,' *J. Assoc. Comp. Mach.* **4**, 329–340.
11 K. W. MORTON (1957). 'A generalization of the antithetic variate technique for evaluating integrals.' *J. Math. and Phys.* **36**, 289–293.
12 S. M. ERMAKOV and V. G. ZOLOTUKHIN (1960). 'Polynomial approximations and the Monte Carlo method' *Teor. Veroyatnost. i Primenen*, **5**, 473–476, translated as *Theor. Prob. and Appl.* **5**, 428–431.
13 F. CERULUS and R. HAGEDORN (1958). 'A Monte Carlo method to calculate multiple phase space integrals.' *Nuovo Cimento*, (X) **9**, *Suppl.* N2, 646–677.
14 N. MANTEL (1953). 'An extension of the Buffon needle problem.' *Ann. Math. Statist.* **24**, 674–677.
15 A. HALL (see chapter 1, reference 4).
16 B. C. KAHAN (1961). 'A practical demonstration of a needle experiment designed to give a number of concurrent estimates of π.' *J. Roy. Statist. Soc.* (A) **124**, 227–239.

Chapter 6

1 H. F. TROTTER and J. W. TUKEY (1956). 'Conditional Monte Carlo for normal samples.' *Symposium on Monte Carlo methods*, ed. H. A. MEYER, 64–79. New York: Wiley.
2 J. M. HAMMERSLEY (1956). 'Conditional Monte Carlo.' *J. Assoc. Comp. Mach.* **3**, 73–76.
3 J. G. WENDEL (1957). 'Groups and conditional Monte Carlo.' *Ann. Math. Statist.* **28**, 1048–1052.

4   H. J. ARNOLD, B. D. BUCHER, H. F. TROTTER and J. W. TUKEY (1956). 'Monte Carlo techniques in a complex problem about normal samples.' *Symposium on Monte Carlo methods*, ed. H. A. MEYER, 80–88. New York: Wiley.

Chapter 7

1   J. H. CURTISS (1956). 'A theoretical comparison of the efficiencies of two classical methods and a Monte Carlo method for computing one component of the solution of a set of linear algebraic equations.' *Symposium on Monte Carlo methods*, ed. H. A. MEYER, 191–233. New York: Wiley.

2   G. E. FORSYTHE and R. A. LEIBLER (1950). 'Matrix inversion by a Monte Carlo method.' *Math. Tabs. Aids Comput.* **4**, 127–129.

3   W. WASOW (1952). 'A note on the inversion of matrices by random walks.' *Math. Tab. Aids Comput.* **6**, 78–81.

4   J. H. HALTON (1962). 'Sequential Monte Carlo.' *Proc. Camb. phil. Soc.* **58**, 57–78.

5   R. E. CUTKOSKY (1951). 'A Monte Carlo method for solving a class of integral equations.' *J. Res. Nat. Bur. Stand.* **47**, 113–115.

6   E. S. PAGE (1954). 'The Monte Carlo solution of some integral equations.' *Proc. Camb. phil. Soc.* **50**, 414–425.

7   B. SPINRAD, G. GOERTZEL and W. SNYDER (1951). 'An alignment chart for Monte Carlo solution of the transport problem.' *Nat. Bur. Stand. Appl. Math. Ser.* **12**, 4–5.

8   G. E. ALBERT (1956). 'A general theory of stochastic estimates of the Neumann series for the solution of certain Fredholm integral equations and related series.' *Symposium on Monte Carlo methods*, ed. H. A. MEYER, 37–46. New York: Wiley.

9   R. COURANT, K. FRIEDRICHS and H. LEWY (1928). 'Über die partiellen Differenzengleichungen der mathematischen Physik.' *Math. Annalen*, **100**, 32–74.

10 J. H. CURTISS (1949). 'Sampling methods applied to differential and difference equations.' *Proc. Seminar on Scientific Computation.* New York: IBM Corporation.

11 M. E. MULLER (1956). 'Some continuous Monte Carlo methods for the Dirichlet problem.' *Ann. Math. Statist.* **27**, 569–589.

12 V. S. VLADIMIROV (1956). 'On the application of the Monte Carlo methods for obtaining the lowest characteristic number and the corresponding eigenfunction for a linear integral equation.' *Teor. Veroyatnost. i Prim.* **1**, 113–130 translated as *Theory of Prob. and Appl.* **1**, 101–116.

13 R. FORTET (1952). 'On the estimation of an eigenvalue by an additive functional of a stochastic process, with special reference to the Kac-Donsker process.' *J. Res. Nat. Bur. Stand.* **48**, 68–75.

14 M. LOÈVE (1960). *Probability Theory* (2nd. Ed.) van Nostrand, Princeton.

15 Z. CIESIELSKI (1961). 'Hölder conditions for realizations of Gaussian processes.' *Trans. Amer. Math. Soc.* **99**, 403–413.

16 M. D. DONSKER and M. KAC (1950). 'A sampling method for determining the lowest eigenvalue and principal eigenfunction of Schrödinger's equation.' *J. Res. Nat. Bur. Stand.* **44**, 551–557.

17 J. L. DOOB (1953). *Stochastic processes.* New York: Wiley.

18 M. KAC (1949). 'On the distribution of certain Wiener functionals.' *Trans. Amer. Math. Soc.* **65**, 1–13.

19 M. KAC (1959). *Probability and related topics in physical sciences.* New York: Interscience.

20 W. WASOW (1951). 'Random walks and the eigenvalues of elliptic differential equations.' *J. Res. Nat. Bur. Stand.* **46**, 65–73.

Chapter 8

1 B. DAVISON (1957). *Neutron Transport Theory.* Oxford Univ. Press.

2 H. KAHN (1956). 'Use of different Monte Carlo sampling techniques.' *Symposium on Monte Carlo Methods*, ed. H. A. MEYER, 146–190. New York: Wiley.

3 H. KAHN and T. E. HARRIS (1951). 'Estimation of particle transmission by random sampling.' *Nat. Bur. Stand. Appl. Math. Ser.* **12**, 27–30.

4 M. J. BERGER (1955). 'Reflection and transmission of gamma radiation by barriers: Monte Carlo calculation by a collision-density method.' *J. Res. Nat. Bur. Stand.* **55**, 343–350.

5 M. J. BERGER and J. DOGGETT (1956). 'Reflection and transmission of gamma radiation by barriers: semianalytic Monte Carlo calculation.' *J. Res. Nat. Bur. Stand.* **56**, 89–98.

6 I. C. PULL (1962). 'Special techniques of the Monte Carlo method.' Chap. 33 of *Numerical Solution of Ordinary and Partial Differential Equations*, ed. L. FOX, 442–457. Oxford: Pergamon Press.

7 K. W. MORTON (1956). 'Criticality calculations by Monte Carlo methods.' *United Kingdom Atomic Energy Research Establishment Report T/R 1903.* Harwell.

8 J. M. HAMMERSLEY and K. W. MORTON (1954). 'Poor man's Monte Carlo.' *J. Roy. Statist. Soc.* (B) **16**, 23–38.

9 K. M. CASE, F. DE HOFFMAN and G. PLACZEK (1953). *Introduction to the theory of neutron diffusion.* Washington D.C.: U.S. Government Printing Office.

Chapter 9

1 W. FELLER (1950). *An introduction to probability theory and its applications.* Vol. I. New York: Wiley.

2 K. L. CHUNG (1960). *Markov chains with stationary transition probabilities.* Berlin: Springer.

3 D. TER HAAR (1960). *Elements of statistical mechanics.* New York: Holt, Rinehart and Winston.

4 A. I. KHINCHIN (1949). *Mathematical foundations of statistical mechanics.* New York: Dover Press.

5 N. METROPOLIS, A. W. ROSENBLUTH, M. N. ROSENBLUTH, A. H. TELLER and E. TELLER (1953). 'Equations of state calculations by fast computing machines.' *J. Chem. Phys.* **21**, 1087–1092.

6 W. W. WOOD and F. R. PARKER (1957). 'Monte Carlo equation of state of molecules interacting with Lennard-Jones potential.' *J. Chem. Phys.* **27**, 720–733.

7 A. W. ROSENBLUTH and M. N. ROSENBLUTH (1954). 'Further results in Monte Carlo equations of state.' *J. Chem. Phys.* **22**, 881–884.

8 Z. W. SALSBURG, J. D. JACOBSON, W. FICKETT and W. W. WOOD (1959). 'Application of the Monte Carlo method to the lattice-gas model.' *J. Chem. Phys.* **30**, 65–72.

9 W. W. WOOD and J. D. JACOBSON (1957). 'Preliminary results from a recalculation of the Monte Carlo equation of state of hard spheres.' *J. Chem. Phys.* **27**, 1207–1208.

10 I. Z. FISHER (1960). 'Applications of the Monte Carlo method in statistical physics.' *Soviet Physics, Uspekhi*, **2**, 783–796.

11 B. J. ALDER and T. E. WAINWRIGHT (1957, 1959 and 1960). 'Studies in molecular dynamics.' *J. Chem. Phys.* **27**, 1208–1209; **31**, 459–466; **33**, 1439–1451.

12 G. F. NEWELL and E. W. MONTROLL (1953). 'On the theory of the Ising model of ferromagnetism.' *Rev. Mod. Phys.* **25**, 353–389.

13 L. D. FOSDICK (1959). 'Calculation of order parameters in a binary alloy by the Monte Carlo method.' *Phys. Rev.* **116**, 565–573.

14 J. R. EHRMAN, L. D. FOSDICK and D. C. HANDSCOMB (1960). 'Computation of order parameters in an Ising lattice by the Monte Carlo method.' *J. Mathematical Phys.* **1**, 547–558.

15 L. D. FOSDICK (1963). 'Monte Carlo calculations on the Ising lattice.' *Methods in Computational Physics*, **1**, 245–280.

16 L. GUTTMAN (1961). 'Monte Carlo computations on the Ising model. The body-centred cubic lattice.' *J. Chem. Phys.* **34**, 1024–1036.

17 D. C. HANDSCOMB (1962). 'The Monte Carlo method in quantum statistical mechanics.' *Proc. Camb. phil. Soc.* **58**, 594–598.

18 D. C. HANDSCOMB (1964). 'A Monte Carlo method applied to the Heisenberg ferromagnet.' *Proc. Camb. phil Soc.* **60**, 115–122.

Chapter 10

1 J. M. HAMMERSLEY (1957). 'Percolation processes. II. The connective constant.' *Proc. Camb. phil. Soc.* **53**, 642–645.

2 J. M. HAMMERSLEY and D. J. A. WELSH (1962). 'Further results on the rate of convergence to the connective constant of the hypercubical lattice.' *Oxford Quart. J. Math.* (2) **13**, 108–110.

3 F. T. WALL (1953). 'Mean dimensions of rubber-like polymer molecules.' *J. Chem. Phys.* **21**, 1914–1919.

4 E. W. MONTROLL (1950). 'Markov chains and excluded volume effect in polymer chains.' *J. Chem. Phys.* **18**, 734–743.

5 F. T. WALL and L. A. HILLER (1954). 'Properties of macromolecules in solution.' *Ann. Rev. Phys. Chem.* **5**, 267–290.

6 J. J. HERMANS (1957). 'High polymers in solution.' *Ann. Rev. Phys. Chem.* **8**, 179–198.

7 E. F. CASASSA (1960). 'Polymer solutions.' *Ann. Rev. Phys. Chem.* **11**, 477–500.

8 R. S. LEHMAN and G. H. WEISS (1958). 'A study of the restricted random walk.' *J. Soc. Indust. Appl. Math.* **6**, 257–278.

9 J. M. HAMMERSLEY (1963). 'Long-chain polymers and self avoiding random walks.' *Sankhyā*, (A) **25**, 29–38, 269–272.

10 M. A. D. FLUENDY and E. B. SMITH (1962). 'The application of Monte Carlo methods to physicochemical problems.' *Quart. Rev. London Chem. Soc.* **16**, 241–266.

11 G. W. KING (1951). 'Stochastic methods in statistical mechanics.' *Nat. Bur. Stand. Appl. Math. Ser.* **12**, 12–18.

12 F. T. WALL, L. A. HILLER, D. J. WHEELER and W. F. ATCHISON (1954 and 1955). 'Statistical computation of mean dimensions of macromolecules. I–III' *J. Chem. Phys.* **22**, 1036–1041; **23**, 913–921, 2314–2321.

13 F. T. WALL, R. J. RUBIN and L. M. ISAACSON (1957). 'Improved statistical method for computing mean dimensions of polymer molecules.' *J. Chem. Phys.* **27**, 186–188.

14 J. M. HAMMERSLEY and K. W. MORTON (see chapter 8, reference 8).

15 M. N. ROSENBLUTH and A. W. ROSENBLUTH (1955). 'Monte Carlo calculation of the average extension of molecular chains.' *J. Chem. Phys.* **23**, 356–359.

16 P. J. MARCER (1960). *Further investigations of the mean dimensions of non-intersecting chains on simple lattices.* Thesis, Oxford.

17 F. T. WALL and J. J. ERPENBECK (1959). 'New method for the statistical computation of polymer dimensions.' *J. Chem. Phys.* **30**, 634–637.

18 F. T. WALL and J. J. ERPENBECK (1959). 'Statistical computation of radii of gyration and mean internal dimensions of polymer molecules.' *J. Chem. Phys.* **30**, 637–640.

19 F. T. WALL and J. MAZUR (1961). 'Statistical thermodynamics of coiling-type polymers.' *Ann. New York Acad. Sci.* **89**, 608–619.

20 F. T. WALL, S. WINDWER and P. J. GANS (1962). 'Monte Carlo procedures for generation of non-intersecting chains.' *J. Chem. Phys.* **37**, 1461–1465.

21 F. T. WALL, S. WINDWER and P. J. GANS (1963). 'Monte Carlo methods applied to configurations of flexible polymer molecules.' *Methods in Computational Physics*, **1**, 217–243.

22   M. A. D. FLUENDY (1962). *The kinetics of intramolecular reactions in solution*. Thesis, Oxford.

Chapter 11

1   S. R. BROADBENT and J. M. HAMMERSLEY (1957). 'Percolation processes. I. Crystals and mazes.' *Proc. Camb. phil. Soc.* **53**, 629–641.
2   H. L. FRISCH and J. M. HAMMERSLEY (1963). 'Percolation processes and related topics.' *J. Soc. Indust. Appl. Math.* **11**, 894–918.
3   J. M. HAMMERSLEY (1963). 'A Monte Carlo solution of percolation in the cubic crystal.' *Methods in Computational Physics*, **1**, 281–298.
4   H. L. FRISCH, S. B. GORDON, J. M. HAMMERSLEY and V. A. VYSSOTSKY (1962). 'Monte Carlo solution of bond percolation processes in various crystal lattices.' *Bell Syst. Tech. J.* **41**, 909–920.
5   H. L. FRISCH, E. SONNENBLICK, V. A. VYSSOTSKY and J. M. HAMMERSLEY (1961). 'Critical percolation probabilities (site problem).' *Phys. Rev.* **124**, 1021–1022.
6   H. L. FRISCH, J. M. HAMMERSLEY and D. J. A. WELSH (1962). 'Monte Carlo estimates of percolation probabilities for various lattices.' *Phys. Rev.* **126**, 949–951.
7   V. A. VYSSOTSKY, S. B. GORDON, H. L. FRISCH and J. M. HAMMERSLEY (1961). 'Critical percolation probabilities (bond problem).' *Phys. Rev.* **123**, 1566–1567.
8   P. DEAN (1963). 'A new Monte Carlo method for percolation problems on a lattice.' *Proc. Camb. phil. Soc.* **59**, 397–410.

Chapter 12

1   H. C. THACHER (1960). 'Introductory remarks on numerical properties of functions of more than one independent variable.' *Ann. New York Acad. Sci.* **86**, 679–681.
2   J. M. HAMMERSLEY (1960). 'Monte Carlo methods for solving multivariable problems.' *Ann. New York Acad. Sci.* **86**, 844–874.
3   C. D. ALLEN (1959). 'A method for the reduction of empirical multivariable functions.' *Computer J.* **1**, 196–200.
4   C. B. TOMPKINS (1956). 'Machine attacks on problems whose variables are permutations.' *Proc. Symp. Appl. Math.* **6**, 195–211.

Further References

The following references are not specifically mentioned in the main text of this book; but it seems useful to collect them together for bibliographical reasons, even if such a bibliography is limited by our own knowledge and is necessarily incomplete. We have classified them under various headings according to their principal contents; but this classification is rather rough and ready and makes little allowance for papers covering several different topics.

General descriptions of the Monte Carlo method

BOUCKHAERT, L. (1956). 'Les méthodes de Monte Carlo.' *Rev. Questions Sci.* (5) **17**, 344–359.

BUSLENKO, N. P., and SREIDER, JU. A. (1961). *The Monte Carlo Method and how it is carried out on Digital Computers.* Moscow: Gosudarstv.

CLARK, C. E. (1960). 'The utility of statistics of random numbers.' *J. Oper. Res. Soc. Amer.* **8**, 185–195.

CSEKE, V. (1961). 'Sur la méthode Monte Carlo.' *Gaz. Mat. Fiz. Ser.* (A) **13** (66), 352–360.

DAHLQUIST, G. (1954). 'The Monte Carlo method.' *Nordisk Mat. Tidskr.* **2**, 27–43, 80.

DUPAC, V. (1962). 'Monte Carlo methods.' *Apl. Mat.* **7**, 1–20.

ENGELER, E. (1958). 'Uber die Monte-Carlo-methode.' *Mitt. Verein. Schweiz. Versich-Math.* **58**, 67–76.

FRANCKX, E. (1956). 'La méthode de Monte Carlo.' *Assoc. Actuar. Belges Bull.* **58**, 89–101.

ULAM, S. (1951). 'On the Monte Carlo method.' *Proc. 2nd Symp. Large-scale Digital Calculating Machinery*, 207–212.

Physical devices for generating random numbers

GOLENKO, D. I., and SMIRJAGIN, V. P. (1960). 'A source of random numbers which are equidistributed in [0,1].' *Magyar Tud. Akad. Mat. Kutato Int. Közl.* **5**, 241–253.

HAMMERSLEY, J. M. (1950). 'Electronic computers and the analysis of stochastic processes.' *Math. Tab. Aids Comput.* **4**, 56–57.

HAVEL, J. (1961). 'An electronic generator of random sequences.' *Trans. 2nd Prague Conf. Information Theory*, 219–225. New York: Academic Press.

HIRAI, H., and MIKAMI, T. (1960). 'Design of random walker for Monte Carlo method. Electronic device.' *J. Inst. Polytech. Osaka City Univ.* (A) **11**, 23–38.

ISADA, M., and IKEDA, H. (1956). 'Random number generator.' *Ann. Inst. Statist. Math. Tokyo*, **8**, 119–126.

ISAKSSON, H. (1959). 'A generator of random numbers.' *Teleteknik*, **3**, 25–40.

SUGIYAMA, H. and MIYATAKE, O. (1959). 'Design of random walker for Monte Carlo method.' *J. Inst. Polytech. Osaka City Univ.* (A) **10**, 35–41.

Generation and properties of pseudorandom numbers

BARNETT, V. D. (1962). 'The behavior of pseudo-random sequences generated on computers by the multiplicative congruential method.' *Math. Comp.* **16**, 63–69.

BASS, J. (1960). 'Nombres aléatoire, suites arithmetiques, méthode de Monte Carlo.' *Publ. Inst. Statist. Univ. Paris*, **9**.

BASS, J., and GUILLOUD, J. (1958). 'Méthode de Monte Carlo et suites uniformement denses.' *Chiffres*, **1**, 149–156.

BERTRANDIAS, J. P. (1960). 'Calcul d'une integrale au moyen de la suite $X_n = A_n$. Evaluation de l'erreur.' *Publ. Inst. Statist. Univ. Paris*, **9**, 335–357.

CERTAINE, J. (1958). 'On sequences of pseudo-random numbers of maximal length.' *J. Assoc. Comp. Mach.* **5**, 353–356.

COVEYOU, R. R. (1960). 'Serial correlation in the generation of pseudo-random numbers.' *J. Assoc. Comp. Mach.* **7**, 72–74.

DUPARC, H. J. A., and PEREMANS, W. (1952). 'Enige methods om random-numbers te maken. II.' *Math. Centrum. Rapp.* Z.W. 1952–013. Amsterdam.

DUPARC, H. J. A., LEKKERKERKER, C. G., and PEREMANS, W. (1953). 'Reduced sequences of integers and pseudo-random numbers.' *Math. Centrum. Rapp.* Z.W. 1953–002. Amsterdam.

FRANKLIN, J. N. (1958). 'On the equidistribution of pseudo-random numbers.' *Quart. Appl. Math.* **16**, 183–188.

HAMMER, P. C. (1951). 'The mid-square method of generating digits.' *Nat. Bur. Stand. Appl. Math. Ser.* **12**, 33.

HOERNER, S. VON (1957). 'Herstellung von Zufallszahlen auf Rechen-automaten.' *Z. Angew. Math. Phys.* **8**, 26–52.

HORTON, H. B., and SMITH, R. T. (1949). 'A direct method for producing random digits in any number system.' *Ann. Math. Statist.* **20**, 82–90.

LINIGER, W. (1961). 'On a method by D. H. Lehmer for the generation of pseudo-random numbers.' *Numerische Math.* **3**, 265–270.

MOSHMAN, J. (1954). 'The generation of pseudo-random numbers on a decimal calculator.' *J. Assoc. Comp. Mach.* **1**, 88–91.

PAGE, E. S. (1959). 'Pseudo-random elements for computers.' *Appl. Statist.* **8**, 124–131.

PEACH, P. (1961). 'Bias in pseudo-random numbers.' *J. Amer. Statist. Assoc.* **56**, 610–618.

ROTENBERG, A. (1960). 'A new pseudo-random number generator.' *J. Assoc. Comp. Mach.* **7**, 75–77.

SOBOL, I. M. (1958). 'Pseudo-random numbers for the machine Strela.' *Teor. Veroyatnost. i Primenen.* **3**, 205–211; translated *Theor. Prob. and Appl.* **3**, 192–197.

SPENSER, G. (1955). 'Random numbers and their generation.' *Computers and Automation*, **4**, no. 3, 10–11, 23.

TOCHER, K. D. (1954). 'The application of automatic computers to sampling experiments.' *J. Roy. Statist. Soc.* (B) **16**, 39–61.

VOTAW, D. F., and RAFFERTY, J. A. (1951). 'High speed sampling.' *Math. Tab. Aids Comput.* **5**, 1–8.

WALSH, J. E. (1949). 'Concerning compound randomization in the binary system.' *Ann. Math. Statist.* **20**, 580–589.

YAMADA, S. (1960/1961). 'On the period of pseudo-random numbers generated by Lehmer's congruential method.' *J. Oper. Res. Soc. Japan*, **3**, 113–123.

Tests of random numbers

BUTCHER, J. C. (1961). 'A partition test for pseudo-random numbers.' *Math. Comp.* **15**, 198–199.

DODD, E. L. (1942). 'Certain tests for randomness applied to data grouped into small sets.' *Econometrica*, **10**, 249–257.

FISSER, H. (1961). 'Some tests applied to pseudo-random numbers generated by v. Hoerner's rule.' *Numerische Math.* **3**, 247–249.

GREEN, B. F., SMITH, J. E. K., and KLEM, L. (1959). 'Empirical tests of an additive random number generator.' *J. Assoc. Comp. Mach.* **6**, 527–537.

GRUENBERGER, F. (1950). 'Tests of random digits.' *Math. Tab. Aids Comput.* **4**, 244–245.

GRUENBERGER, F., and MARK, A. M. (1951). 'The d^2 test of random digits.' *Math. Tab. Aids Comput.* **5**, 109–110.

KENDALL, M. G., and BABINGTON SMITH, B. (1938). 'Randomness and random sampling numbers.' *J. Roy. Statist. Soc.* (A) **101**, 147–166.

KENDALL, M. G., and BABINGTON SMITH, B. (1939). 'Second paper on random sampling numbers.' *J. Roy. Statist. Soc.* (B) **6**, 51–61.

MOORE, P. G. (1953). 'A sequential test for randomness.' *Biometrika*, **40**, 111–115.

YOUNG, L. C. (1941). 'On randomness in ordered sequences.' *Ann. Math. Statist.* **12**, 293–300.

Generation of random variables with prescribed distributions

AZORIN, F., and WOLD, H. (1950). 'Product sums and modulus sums of H. Wold's normal deviates.' *Trabajos Estadist.* **1**, 5–28.

CLARK, C. E., and HOLZ, B. W. (1960). *Exponentially distributed random numbers*. Baltimore: Johns Hopkins Press.

GOLENKO, D. I. (1959). 'Generation of random numbers with arbitrary distribution law.' *Vycisl. Mat.* 5, 83–92.

HICKS, J. S., and WHEELING, R. F. (1959). 'An efficient method for generating uniformly distributed points on the surface of an n-dimensional sphere.' *Comm. Assoc. Comp. Mach.* 2, no. 4, 17–19.

MULLER, M. E. (1958). 'An inverse method for the generation of random normal deviates on large-scale computers.' *Math. Tab. Aids Comput.* 12, 167–174.

MULLER, M. E. (1959). 'A note on a method for generating points uniformly on N-dimensional spheres.' *Comm. Assoc. Comp. Mach.* 2, no. 4, 19–20.

QUENOUILLE, M. H. (1959). 'Tables of random observations from standard distributions.' *Biometrika*, 46, 178–202.

RAO, C. R. (1961). 'Generation of random permutations of a given number of elements using random sampling numbers.' *Sankhyā* (A) 23, 305–307.

SCHEUER, E. M., and STOLLER, D. S. (1962). 'On the generation of normal random vectors.' *Technometrics*, 4, 278–281.

Simulation

ALLRED, J. C., and NEWHOUSE, A. (1958). 'Applications of the Monte Carlo method to architectural acoustics.' *J. Acoust. Soc. Amer.* 30, 1–3.

BARTLETT, M. S. (1953). 'Stochastic processes or the statistic of change.' *Appl. Statist.* 2, 44.

BLUMSTEIN, A. (1957). 'A Monte Carlo analysis of the ground-controlled approach system.' *J. Oper Res. Soc. Amer.* 4, 397–408.

BROADBENT, S. R. (1956). 'A measure of dispersion applied to cosmic-ray and other problems.' *Proc. Camb. phil. Soc.* 52, 499–513.

BROTMAN, L., and MINKER, J. (1957). 'Digital simulation of complex traffic problems in communications systems.' *J. Oper. Res. Soc. Amer.* 5, 670–678.

COCHRAN, W. G. (1946). 'Use of IBM equipment in an investigation of the truncated normal problem.' *Proc. Res. Forum*. New York: I.B.M. Corporation.

CURTIN, K. M. (1959) 'A Monte Carlo approach to evaluate multimodel reliability.' *J. Oper. Res. Soc. Amer.* 7, 721–727.

CUSHEN, W. E. (1956). 'Operational gaming in industry.' *Operations Research for Management*, ed. J. F. MCCLOSKEY and J. M. COPPINGER, 2, 358–375. Baltimore: Johns Hopkins Press.

CUSIMANO, G. (1959). 'Il metodo di Montecarlo e le sue applicazione nello statistica industriale.' *Statistica*, 19, 216–229.

DAVIS, D. H. (1960). 'Monte Carlo calculation of molecular flow rates through a cylindrical elbow and pipes of other shapes.' *J. Appl. Phys.* 31, 1169–1176.

GALLIHER, H. P. (1957). 'Monte Carlo simulation studies.' *Report of System Simulation Symposium*, 24–27. New York.

GANI, J., and MORAN, P. A. P. (1955). 'The solution of dam equations by Monte Carlo methods.' *Austral. J. Appl. Sci.* 6, 267–273.

GOETZ, B. E. (1960). 'Monte Carlo solution of waiting line problems.' *Management Technology*, 1, 2–11.

GORN, S. (1959). 'On the mechanical simulation of habit-forming and learning.' *Information and Control*, 2, 226–259.

HAITSMA, A. (1959). 'Monte Carlo and simulation.' *Statistica Neerlandica*, 13, 395–406.

HARLING, J. (1958). 'Simulation techniques in operations research – a review.' *J. Oper. Res. Soc. Amer.* 6, 307–319.

HOFFMAN, J., METROPOLIS, N., and GARDINER, V. (1956). 'Study of tumor cell populations by Monte Carlo methods.' *Science*, 122, 465–466.

HUBBARD, H. B. (1957). 'Monte Carlo applications.' *Second Annual Quality Control Conference on Methods and Management*. Drexel Institute.

JESSOP, W. N. (1956). 'Operational research methods: What are they?' *Oper. Res. Quart.* 7, 49–58.

JESSOP, W. N. (1956). 'Monte Carlo methods and industrial problems.' *Appl. Statist.* 5, 156–165.

JONES, H. G., and LEE, A. M. (1955). 'Monte Carlo methods in heavy industry.' *Oper Res. Quart.* 2, 108–116.

KARR, H. W. (1955). 'A Monte Carlo model of an Air Force type supply system.' *Rand Corporation Report* P-683A.

KELLEY, D. H., and BUXTON, J. N. (1962). 'Montecode – an interpretive program for Monte Carlo simulations.' *Computer J.* 5, 88–93.

KENDALL, D. G. (1950). 'An artificial realization of a simple birth-and-death process.' *J. Roy. Statist. Soc.* (B) 12, 116–119.

KING, G. W. (1952). 'The Monte Carlo method as natural mode of expression in operations research.' *J. Oper. Res. Soc. Amer.* 1, 46–51.

KLIMOV, G. P. (1961). 'Simulation on electronic digital computers of a class of queuing systems.' *Z. Vycisl. Mat. i Mat. Fiz.* 1, 935–940.

KLIMOV, G. P. and ALLEN, G. A. (1961). 'Computer solution of a problem in the theory of queues by a Monte Carlo method.' *Z. Vycisl. Mat. i Mat. Fiz.* 1, 933–935.

MUSK, F. I. (1959). 'A Monte Carlo simulation of a production planning problem.' *Computer J.* 2, 90–94.

NEATE, R., and DACEY, W. J. (1958). 'A simulation of melting shop operations by means of a computer.' *Process Control and Automation.* 5, 264–272.

THOMAS, C. J. (1957). 'The genesis and practice of operational gaming.' *Proc. 1st Internat. Conf. Oper. Res.* 64–81.

TOCHER, K. D., and OWEN, D. G. (1960). 'The automatic programming of simulation.' *Proc. 2nd Internat. Conf. on Oper. Res.* 49–68.

VASONYI, A. (1957). 'Electronic simulation of business operations (the

Monte Carlo Method).' *Second Annual West Coast Management Conference*.

WAGNER, H. M. (1962). *Statistical management of inventory systems*. New York: Wiley.

WALKER, S. H. (1956). 'Applications of the Monte Carlo method in systems analysis.' *Proc. 7th Ann. Nat. Conf. AIIE*. Washington, D.C.

WALSH, J. E. (1963). 'Use of linearised non-linear regression for simulations involving Monte Carlo.' *J. Oper. Res. Soc. Amer.* **11**, 228–235.

YOULE, P. V., TOCHER, K. D., JESSOP, J. W., and MUSK, F. I. (1959). 'Simulation studies of industrial operations.' *J. Roy. Statist. Soc.* (A) **122**, 484–510.

ZIMMERMAN, R. E. (1955). 'Monte Carlo computer war gaming.' *J. Oper. Res. Soc. Amer.* **3**, 120.

ZIMMERMAN, R. E. (1956). 'A Monte Carlo model for military analysis.' *Operations Research for Management*, ed. J. P. MCCLOSKEY and J. W. COPPINGER, **2**, 376–400. Baltimore: Johns Hopkins Press.

Integration

DAVIS, P., and RABINOWITZ, P. (1956). 'Some Monte Carlo experiments in computing multiple integrals.' *Math. Tab. Aids Comput.* **10**, 1–8.

KAHN, H. (1960). 'Multiple quadrature by Monte Carlo methods.' *Mathematical methods for digital computers*, 249–257. New York: Wiley.

PALASTI, I., and RÉNYI, A. (1956). 'Monte Carlo methods as minimax strategies.' *Magyar. Tud. Akad. Mat. Kutato Int. Közl.* **1**, 529–545.

SOBOL, I. M. (1961). 'Evaluation of infinite-dimensional integrals.' *Z. Vycisl. Mat. i Mat. Fiz.* **1**, 917–922.

SOBOL, I. M. (1957). 'Multidimensional integrals and the Monte Carlo method.' *Dokl. Akad. Nauk. SSSR (NS)*, **114**, 706–709.

Variance-reducing techniques

GEL'FAND, I. M., FROLOV, A. S., and CENCOV, N. N. (1958). 'The computation of continuous integrals by the Monte Carlo method.' *Izv. Vyss. Ucebn. Zaved. Matamika*, **5** (6), 32–45.

HAMMERSLEY, J. M. (1962). 'Monte Carlo methods.' *Proc. 7th Conf. on Design of Experiments in Army Research, Development and Testing*, 17–26. Durham: U.S. Army Research Office.

KAHN, H. (1960). 'Multiple quadrature by Monte Carlo methods.' *Mathematical methods for digital computers*, 249–257. New York: Wiley.

KAHN, H., and MARSHALL, A. W. (1953). 'Methods of reducing sample size in Monte Carlo computations.' *J. Oper. Res. Soc. Amer.* **1**, 263–271.

MOSHMAN, J. (1958). 'The application of sequential estimation to computer simulation and Monte Carlo procedures.' *J. Assoc. Comp. Mach.* **5**, 343–352.

Linear operator problems

AKAIKE, H. (1956). 'Monte Carlo method applied to the solution of simultaneous linear equations.' *Ann. Inst. Statist. Math. Tokyo*, **7**, 107–113.

BAUER, W. F. (1958). 'The Monte Carlo method.' *J. Soc. Indust. Appl. Math.* **6**, 438–451.

BAZLEY, N. W., and DAVIS, P. J. (1960). 'Accuracy of Monte Carlo methods in computing finite Markov chains.' *J. Res. Nat. Bur. Stand.* **64B**, 211–215.

CURTISS, J. H. (1953). 'Monte Carlo methods for the iteration of linear operators.' *J. Math. and Phys.* **32**, 209–232.

DUPAC, V. (1956). 'Stochastic numerical methods.' *Casopia Pest. Mat.* **81**, 55–68.

EDMUNDSON, H. P. (1952). 'Monte Carlo matrix inversion and recurrent events.' *Math. Tab. Aids Comput.* **7**, 18–21.

EHRLICH, L. W. (1959). 'Monte Carlo solutions of boundary value problems involving the difference analogue of $\partial^2 u/\partial x^2 + \partial^2 u/\partial y^2 + Ky^{-1}\partial u/\partial y = 0$.' *J. Assoc. Comp. Mach.* **6**, 204–218.

GATESOUPE, M., and GUILLOUD, J. (1961). 'Une application de la méthode de Monte Carlo à la resolution numerique des equations de Fredholm.' *Publ. Inst. Statist. Univ. Paris*, **10**, 243–265.

GOLENKO, D. I. (1959). 'Calculation of the characteristic of certain stochastic processes by the Monte Carlo method.' *Vycisl. Mat.* **5**, 93–108.

HIRAI, H., and MIKAMI, T. (1960). 'Design of random walker for Monte Carlo method. Electronic device.' *J. Inst. Polytech. Osaka City Univ.* (A) **11**, 23–28.

KAC, M. (1951). 'On some connections between probability theory and differential and integral equations.' *Proc. 2nd Berkeley Symp. on Math. Statist. and Probability*, 189–215. Berkeley: Univ. of California Press.

KALOS, M. H. (1962). 'Monte Carlo calculations of the ground state of three- and four-body nuclei.' *Phys. Rev.* **128**, 1791–1795.

KLAHR, C. N. (1960). 'A Monte Carlo method for the solution of elliptic partial differential equations.' *Mathematical methods for digital computers*, 157–164. New York: Wiley.

MOTOO, M. (1959). 'Some evaluations for continuous Monte Carlo method.' *J. Inst. Statist. Math. Tokyo*, **11**, 35–41.

NAGAR, A. L. (1960). 'A Monte Carlo study of alternative simultaneous equation estimators.' *Econometrica*, **28**, 573–590.

OPLER, A. (1951). 'Monte Carlo matrix calculations with punched card machines.' *Math. Tab. Aids Comput.* **5**, 115–120.

OSWALD, F. J. (1960). 'Matrix inversion by Monte Carlo methods.' *Mathematical methods for digital computers*, 78–83. New York: Wiley.

THOMAS, L. H. (1953). 'A comparison of stochastic and direct methods for the solution of some special problems.' *J. Oper. Res. Soc. Amer.* **1**, 181–186.

TODD, J. (1954). 'Experiments in the solution of a differential equation by Monte Carlo methods.' *J. Wash. Acad. Sci.* **44**, 377–381.

TODD, J. (1953). 'Experiments on the inversion of a 16×16 matrix.' *Nat. Bur. Stand. Appl. Mat. Ser.* **29**, 113–115.

VLADIMIROV, V. S., and SOBOL, I. M. (1958). 'Computations of the least eigenvalue of the Peierls equations by the Monte Carlo method.' *Vycisl. Mat.* **3**, 130–137.

WASOW, W. (1951). 'On the mean duration of random walks.' *J. Res. Nat. Bur. Stand.* **46**, 462–471.

WASOW, W. (1951). 'On the duration of random walks.' *Ann. Math. Statist.* **22**, 199–216.

YOWELL, E. C. (1949). 'A Monte Carlo method of solving Laplace's equation.' *Proc. Seminar on Scientific Computation*, 87–91. New York: I.B.M. Corporation.

Nuclear particle transport

BERGER, M. J. (1963). 'Monte Carlo calculation of the penetration and diffusion of fast charged particles.' *Methods in Computational Physics*, **1**, 135–215.

BERNADINI, G., BOOTH, E. T., and LINDENBAUM, S. J. (1952). 'The interactions of high energy nucleons with nuclei. II.' *Phys. Rev.* **88**, 1017–1026.

BUTCHER, J. C. and MESSEL, H. (1958). 'Electron number distribution in electron-photon showers.' *Phys. Rev.* **112**, 2096–2106.

CASHWELL, E. D., and EVERETT, C. J. (1959). *A practical manual on the Monte Carlo method for random walk problems.* New York: Pergamon Press.

CREW, J. E., HILL, R. D., and LAVATELLI, L. S. (1957). 'Monte Carlo calculations of single pion production by pions.' *Phys. Rev.* **106** 1051–1056.

CÜER, P., and COMBE, J. (1954). 'Sur la realization d'une technique de Monte Carlo pour etudier le passage des nucleons de grande energie à travers les noyaux.' *Comptes rendus*, **238**, 1799–1801.

DAVIS, D. H. (1963). 'Critical size calculations for neutron systems by the Monte Carlo method.' *Methods in Computational Physics*, **1**, 67–88.

DEMARCUS, W. C., and NELSON, L. (1951). 'Methods of probabilities in chains applied to particle transmission through matter.' *Nat. Bur. Stand. Appl. Math. Ser.* **12**, 9–11.

FERMI, E., and RICHTMYER, R. D. (1948). 'Note on census-taking in Monte Carlo calculations.' LADC-946. Los Alamos.

FLECK, J. A. (1963). 'The calculation of non-linear radiation transport by a Monte Carlo method.' *Methods in Computational Physics*, **1**, 43–65.

FORTET, R. (1953). 'Applications de la statistique à la physique nucleaire.' *Monografias de ciencia moderna*, **40**. Madrid.

GOAD, W., and JOHNSTONE, R. (1959). 'A Monte Carlo method for criticality problems.' *Nuc. Sci. and Eng.* **5**, 371–375.

GOERTZEL, G. (1949). 'Quota sampling and importance functions.' AECD-2793. Oak Ridge.

GOERTZEL, G. and KALOS, M. H. (1958). 'Monte Carlo methods in transport problems.' *Progress in Nuclear Energy*, **2**, 315–369.

GOLDBERGER, M. L. (1948). 'The interaction of high energy neutrons and heavy nuclei.' *Phys. Rev.* **74**, 1269–1277.

HAMMER, P. C. (1951). 'Calculation of shielding properties of water for high energy neutrons.' *Nat. Bur. Stand. Appl. Math.* **12**, 21–23.

JOHNSTON, R. R. (1963). 'A general Monte Carlo neutronics code.' LAMS-2856, Los Alamos.

KAHN, H. (1950). 'Random sampling (Monte Carlo) techniques in neutron attenuation problems.' *Nucleonics*, **6**, 27–33, 36, 60–65.

METROPOLIS, N., BIVINS, R., STORM, M., MILLER, J. M., FRIEDLANDER, G., and TURKEVICH, A. (1958). 'Monte Carlo calculations on intranuclear cascades.' *Phys. Rev.* **110**, 185–219.

RICHTMYER, R. D. (1961). 'Monte Carlo methods.' *Proc. Symp. Appl. Math.* **11**, 190–205.

SPANIER, J. (1962). 'A unified approach to Monte Carlo methods and an application to a multigroup calculation of absorption rates.' *Soc. Indust. Appl. Math. Rev.* **4**, 115–134.

WILSON, R. R. (1951). 'The range and straggling of high energy electrons.' *Phys. Rev.* **84**, 100–103.

WILSON, R. R. (1952). 'Monte Carlo study of shower production.' *Phys. Rev.* **86**, 261–269.

ZERBY, C. D. (1963). 'A Monte Carlo calculation of the response of gamma-ray scintillation counters.' *Methods in Computational Physics*, **1**, 89–134.

Miscellaneous

CHESNUT, D. A., and SALSBURG, Z. W. (1963). 'Monte Carlo procedure for statistical mechanical calculations in a grand canonical ensemble of lattice systems.' *J. Chem. Phys.* **38**, 2861–2875.

KASTELEYN, P. W. (1963). 'A soluble self-avoiding walk problem.' *Physica*, **29**, 1329–1337.

KESTEN, H. (1963). 'On the number of self-avoiding walks.' *J. Mathematical Phys.* **4**, 960–969.

LEEUW, K. DE, MOORE, R. F., SHANNON, C. E., and SHAPIRO, N. (1956). 'Computability by probabilistic machines.' *Ann. Math. Studies*, **34**, 183–212.

PASTA, J. R., and ULAM, S. (1959). 'Heuristic numerical work in some problems of hydrodynamics.' *Math. Tab. Aids Comput.* **13**, 1–12.

SCOTT, G. D. (1962). 'Radial distribution of the random close packing of equal spheres.' *Nature*, **194**, 956–957.

SUGIYAMA, H., and JOH, K. (1962). 'A numerical procedure for conformal mapping in case of simply, doubly, and multiply connected domains from the viewpoint of the Monte Carlo approach. I.' *Tech. Rep. Osaka Univ.* **12**, 1–9.

TAUB, A. H. (1950). 'A sampling method for solving the equations of compressible flow in a permeable medium.' *Proc. Midwestern Conf. on Fluid Dynamics*, 121–127.

Index

Entries for authors in the index quote pages on which either the author's name appears or one of his publications is cited by its reference number.

Standard error, of binomial para-
meter, 21
of mean, 21
of variance, 21
Standardized normal distributions,
15
Standardized rectangular distri-
bution, 15
State, equations of, 121–2
Markov, 113
Stationary Markov chain, 113
Statistical mechanics, 8, 96, 113–26,
167
Statistical tests of random numbers,
25, 29–31, 161
Stieltjes integral, 11
Stratified sampling, 24, 55–7, 65,
107
Stuart, A., 10, 150
Student, 7
Stochastic matrix, 93, 115
Stoller, D. S., 162
Storage systems, 44
Storm, M., 167
Strides, self-avoiding, 129
Subcritical, 105
Substitutional alloy, 122
Sugiyama, H., 160, 168
Supercritical, 105

Taub, A. H., 168
Taussky, O., 29, 151
Teichroew, D., 39, 151
Telephone system, 3
Teller, A. H., 118, 121, 156
Teller, E., 118, 121, 156
ter Haar, D., 117, 156
Termination probability, 86
Tetrahedral lattice, 127
Thacher, H. C., 142, 158
Theoretical mathematics, 3
Thermal equilibrium, 117
Thomas, C. J., 163
Thomas, L. H., 165
Thomson, M. J., 46, 152

Thomson, W. E., 3, 150
Tocher, K. D., 44, 152, 161, 163, 164
Todd, J., 29, 151, 166
Tompkins, C. B., 149, 158
Transient state, 115
Transition probability, 86, 113
Transition probability, n-step, 114
Travelling salesman problem, 47–8
Trotter, H. F., 80, 83, 84, 153, 154
Tukey, J. W., 61, 80, 83, 84, 153, 154
Turkevich, A., 167

Unbiased estimator, 18
Uncorrelated, 14
Ulam, S., 8, 86, 159, 167

van der Corput, J. C., 33, 151
van der Corput sequences, 33
Variance, 13
Variance, analysis of, 23
Variance-covariance matrix, 16, 19
Variance ratio, 51, 65
Variance-reducing techniques, 5, 8,
22, 55–75, 164
comparison of, 65
Variance, sampling, 18
Vector random variable, 12
Vladimirov, V. S., 94, 154, 166
von Hoerner, S., 160
von Neumann, J., 8, 27, 36, 37, 86,
151
Votaw, D. F., 161
Vyssotsky, 139, 141, 158

Wagner, H. M., 164
Wainwright, T. E., 122, 156
Walk, Pólya, 91, 127
Walker, S. H., 164
Walks in continuous space, 132–3
Walks, self-avoiding, 114, 127–32,
167
Wall, F. T., 128, 129, 131, 132, 157
Walsh, J. E., 161, 164
War-gaming, 44, 162–4
Wasow, W., 87, 90, 96, 154, 155, 166
Water control policies, 43